T0197824

Scientific Freedom

GLOBAL EPISTEMICS

In partnership with the Centre for Global Knowledge Studies (gloknos)

Founding Editor:

Inanna Hamati-Ataya (University of Cambridge)

Editorial Assistants:

Felix Anderl and **Matthew Holmes** (University of Cambridge)

Editorial Review Board:

tinyurl.com/GlobalEpistemics | tinyurl.com/RLIgloknos

Titles in the Series:

Scientific Freedom

The Heart of the Right to Science

Sebastian Porsdam Mann
Helle Porsdam
Maximilian M. Schmid
Péter Vilmos Treit

ROWMAN & LITTLEFIELD
Lanham • Boulder • New York • London

Published by Rowman & Littlefield
An imprint of The Rowman & Littlefield Publishing Group, Inc.
4501 Forbes Boulevard, Suite 200, Lanham, Maryland 20706
www.rowman.com

86-90 Paul Street, London EC2A 4NE, United Kingdom

British Library Cataloguing in Publication Information Available

Library of Congress Cataloging-in-Publication Data

Library of Congress Cataloging-in-Publication Data

Names: Mann, Sebastian Porsdam, author. Title: Scientific freedom : a guide to the right
    to science / Sebastian Porsdam Mann, Maximilian M. Schmid, Péter Vilmos Treit,
    Helle Porsdam.
Description: Lanham : Rowman & Littlefield, [2023] | Series: Global epistemics |
    Includes bibliographical references and index. |
Summary: "In a time of genetic editing, global warming, and a worldwide pandemic, the
    question of how freely science is and should be conducted is one that has significant
    practical consequences. Drawing on rigorous interdisciplinary methods, this book
    develops a model of scientific freedom as a human right"— Provided by publisher.
Identifiers: LCCN 2023038458 (print) | LCCN 2023038459 (ebook) | ISBN
    9781538178379 (cloth) | ISBN 9781538178393 (electronic)
Subjects: LCSH: Science—Social aspects. | Science—Philosophy. | Science—Political
    aspects. | Research—Social aspects. | Research—Philosophy. | Research—Political
    aspects. | Academic freedom. | Human rights.
Classification: LCC Q175.5 .M355 2023 (print) | LCC Q175.5 (ebook) | DDC
    501—dc23/eng/20231016
LC record available at https://lccn.loc.gov/2023038458
LC ebook record available at https://lccn.loc.gov/2023038459

Contents

Series Editor's Note

Scientific freedom is a core principle and value of scientific activity, but its practice, protection, and regulation are complex processes that unfold in a larger societal context in which scientific research and its practical uses may potentially affect the well-being and future development of humanity as a whole. At a time of existential crisis when science and technology are increasingly called on to offer solutions to our ecological predicament, while the capacity to deploy scientific and technological innovations in pursuit of less universal goals becomes more widespread across a variety of more or less regulated social groups and organisations, a nuanced and purposeful reassessment of scientific freedom and its corollary, scientific responsibility, is an especially timely and valuable endeavour.

The authors of this book offer us a novel perspective on scientific freedom that places it at the heart of society's and humanity's right to participate in and benefit from the advances of science and technology. This 'right to science' is one of the least well-known of the 'universal cultural rights' recognised by states in international law, and is consequently often neglected in public, political and academic discussions of human rights, community development or protection from technologically mediated harm. By demonstrating that scientific freedom is a constitutive element of the right to science, this book not only highlights the properly social nature and effects of the exercise of this professional freedom beyond the concerns of its immediate practitioners, but also offers a clear doctrinal-legal framework for its protection and just regulation for the benefit of society as a whole.

The practical applications and implications of this perspective are wide and far-reaching, and I hope that the pioneering work presented in the following pages will encourage scientists, academics, rights activists, legal experts and responsible policymakers to translate it into a progressive program of action.

Inanna Hamati-Ataya
(Cambridge, 30 June 2023)

Artificial Intelligence

As we venture further into the age of information and innovation, science and technology have become inextricably woven into the fabric of our everyday lives. Advancements in such fields as artificial intelligence, biotechnology and environmental science have immense potential to improve our world. Yet their benefits are unevenly distributed, and the very same power that promises such progress also carries with it the risk of intentional or accidental harm and misuse. In today's interconnected world, these issues transcend borders and call for an in-time, concerted effort at the global level to address them effectively. Solutions must be found that foster a continuous stream of scientific discovery and ensure that the pursuit of knowledge and innovation is guided by principles aimed at benefiting and respecting the rights and interests of everyone concerned.

This book explores these challenges and opportunities through the lens of human rights, specifically the right to science, and scientific freedom.[1] Drafted and continually developed by individuals from a wide range of professional, political and ethnic backgrounds, international human rights law, despite its flaws, is the closest thing we have to a truly global normative discourse reflecting common positions on key legal and ethical standards.[2]

The right to science was established by a collaborative effort of representatives from Latin America, China, France, Lebanon, the United States and the USSR in the aftermath of the Second World War. The result of their labours is a vision of scientific progress, freedom, and responsibility that is more pertinent today than perhaps ever before. Their vision underscores the importance of scientific progress for both material well-being and human development. It emphasizes the necessity of protecting and promoting science by providing extensive freedom to researchers, while also establishing well-defined guidelines to limit activities that may harm the common public interest.

To illustrate the practical relevance of these abstractions, we begin our exploration of scientific freedom with a case study on current advancements in artificial intelligence (AI). Supported by the kind of blue-skies research funding we argue is crucial for guaranteeing both scientific freedom and scientific progress, these tools possess tremendous potential to benefit individuals worldwide, primarily due to their democratising impact on accessing, generating, and manipulating information. This democratisation enhances scientific freedom by enabling a wider range of individuals to engage more extensively with science. However, accompanying this positive potential is a significant threat of deliberate and accidental misuse.

AI, once merely a figment of the imagination and the central theme of science-fiction films since the 1930s, has now become an inseparable part of our daily lives, powering applications as diverse as navigation aids, recommendation systems, and medical diagnostics. Though perhaps now taken for granted, this transition from science fiction to reality was by no means a foregone conclusion. Like many areas of science, the history of AI is full of false starts, promising avenues turned to detours, and unexpected successes.

NEURAL NETS AND BLUE-SKIES FUNDING

One of the most fruitful current approaches to AI is known as deep learning. Deep learning has its origins in electrical circuits inspired by human brain function and architecture, known as Artificial Neural Networks (ANNs). Like neurons, these networks consist of interconnected nodes that transmit electrical signals, activating when they surpass a voltage threshold. In ANNs, adaptation occurs as the network adjusts connection strengths based on repeated patterns in input data, effectively 'learning' from experience.[3]

Despite its current successes, deep learning represents just one of many once promising approaches to AI. Indeed, from the 1950s to the 1990s, a prominent research direction attempted to develop AI not by implicit learning modeled on neural architecture, but instead by encoding explicit logic, search, and reasoning processes. This approach, known as symbolic AI, enjoyed early successes, including applications capable of proving mathematical theorems and playing Checkers.

While symbolic AI advanced forward, ANNs had hit a series of roadblocks. Difficulties related to algorithmic architecture, inadequate computer hardware, and lack of available data made it difficult or impossible to scale ANNs beyond a certain size. When these difficulties turned out to be harder to tackle than expected, the field 'mostly stood still for two decades.'[4]

With no further practical applications in sight, funding from industry and applied government grants dried up. Yet funds earmarked for curiosity-driven, fundamental research allowed ANN researchers—notably Geoffrey Hinton, Yoshua Bengio and Yann Le Cun—to keep exploring and developing their interest in neural networks.[5] This unbridled support ultimately enabled the discovery of key theoretical and hardware advances in the following decades. These advances, combined with massive datasets, serve as the foundation for much of modern AI.

A key example of recent progress is generative AI programs which can produce text, image, video, sounds or other patterned information in response to input. Generative AI systems focused on text are called 'large-scale language models' (LLMs) and include software such as OpenAI's GPT series.

ACCESS TO AND PARTICIPATION IN SCIENTIFIC PROGRESS AND ITS APPLICATIONS

On November 30, 2022, San Francisco–based AI company OpenAI released ChatGPT, a powerful AI chatbot capable of rapidly generating coherent, human-like text in response to user inputs. Its impressive capacities attracted a million users within five days and over a billion monthly users by February 2023, making it the fastest-growing consumer application in history.[6]

ChatGPT, its successor GPT-4, released in March 2023, and competing models such as Google's Bard, Meta's LLaMA, as well as several open-source initiatives, are trained on immense datasets consisting of billions of words of natural language text largely sourced from the internet and complemented by layers of human feedback and reinforcement. They function by predicting the most likely next word in a series of words, based on the statistical patterns and linguistic forms contained in their datasets.[7] This deceptively simple mechanism underlies a surprisingly powerful general ability to retrieve, identify, summarize, translate, predict, organize and suggest relevant information.

Among the countless impressive demonstrations of current LLMs' capacities are their ability to pass college and professional exams,[8] translate text into different languages, including code for use in programming[9] and their use in creating video games[10] and writing novels.[11] Scholars have identified numerous potentially positive impacts of the technology in healthcare, research, law and education, but also significantly disruptive potential in areas including journalism, publishing, content generation, marketing, communications, education, literature and politics.[12]

While GPT-4 is currently only accessible through a monthly subscription, ChatGPT and several other models are free to use by anyone with an internet connection. Because they accept natural language as input, they are much more intuitive to use than many other AI systems. To the extent that they are widely accessible and user-friendly for nonexperts, LLM technologies have enormous potential to democratize access to and participation in science by enhancing the powers of individuals to generate and manipulate information.

For example, GPT-4 can be used to simplify complex scientific information, translating it into more accessible language. This can be used to enhance both personal education and science communication. It can also effortlessly translate scientific data across languages, empowering non-English speakers to contribute to the global scientific conversation. It can assist researchers with various tasks, such as literature reviews, hypothesis generation and data analysis. By accelerating research processes and fostering collaboration, GPT-4 enables a more diverse group of individuals to participate in scientific activities. The powerful potential of this technology is illustrated by the fact that this paragraph was itself written by GPT-4, lightly edited by the authors for length and style. Another illustration of the power of generative AI more generally is provided by the cover art for this book: Artist Teresa Flanderka brought to life an image generated by the authors using MidJourney, based on the title of this book.[13]

The unprecedented accessibility and utility of LLMs like GPT-4 raise important questions about the interplay between scientific freedom and human rights. While these technologies hold tremendous potential for advancing knowledge and fostering inclusivity, they also give rise to a range of potential challenges related to the reliability of and responsibility for outputs, their impact on work and employment, inherent biases and environmental sustainability, as well as the dual-use nature of this technology.[14]

HARMS AND DUAL USE

Because LLMs like ChatGPT function by predicting the statistically most probable subsequent word in a sentence, they are optimised for statistical patterns and linguistic form rather than for truth or reliability of outputs. Though ongoing work is attempting to ameliorate this issue by combining LLMs with knowledge databases and internet searches, the models themselves are currently susceptible to 'hallucinations,' which means that they may confidently provide factually incorrect responses.[15]

This is hardly surprising, given the statistical nature of the software, and considering that every step of the pipeline that generates the software contains inherent biases.[16] Much of the data on which AI is trained originates from internet text disproportionately written by young, white males, which is then labeled and annotated by another group of unrepresentative individuals.[17] This data is crunched by algorithms developed by software engineers with their own unique backgrounds and prejudices, and overseen and shipped by executives with yet further biases and perspectives. Though they contain content filters aimed at preventing harmful uses, these can be circumvented.[18] This potential for both good and bad uses is known as 'dual use' and is demonstrated by the following examples.

DeepMind's AlphaFold is a groundbreaking protein-folding prediction software which uses deep learning and other AI-based technologies to accurately predict and generate protein structures, revolutionising structural bioinformatics.[19] Validated by structural biologists, AlphaFold's predictions save researchers time and resources, uncover previously overlooked ancient relationships in protein evolution and expand our knowledge of protein structures and functions. Given the fundamental role of proteins in nearly all biological functions, AlphaFold and similar generative models have immense potential to facilitate scientific progress in numerous fields, including but not limited to drug discovery.[20]

Another group of scientists adapted a generative AI model to predict and design safe chemical structures for scientific study. However, by simply inverting the scoring method they used to train the model, the scientists coaxed the model to design forty thousand highly toxic chemicals within six hours, including novel chemical warfare agents that were more potent than their predecessors were. 'By inverting the use of our machine learning models, we had transformed our innocuous generative model from a helpful tool of medicine to a generator of likely deadly molecules.'[21]

LIMITING SCIENCE AND ITS APPLICATIONS

The example of the inverted chemical model illustrates the need to address potentially harmful uses of powerful technologies. Indeed, scientific freedom entails not only the pursuit of knowledge and innovation, but also the capacity for introspection and reflection on the potential consequences of the technology being developed. In this context, exercising restraint and limiting the application of technology can be seen as an expression of freedom and responsibility.

Less than six months following the release of ChatGPT, the rapid evolution of AI technology, accompanied by numerous examples of dual-use potential, had triggered increasingly dire warnings from some of the most prominent figures in the AI field and beyond. Among these warnings was an open letter calling for a moratorium on LLM development beyond the current GPT-4 standard for an initial period of six months, signed by thousands of leading intellectuals, including Yoshua Bengio, one of the founders of the underlying technology.[22] The open-letter recommended policy measures such as mandating robust third-party auditing and certification, regulating access to computational power, establishing capable AI agencies at the national level, assigning liability for AI-caused harms, and introducing measures to prevent and track AI model leaks. It also called for the expansion of technical AI safety research funding, and the development of standards for identifying and managing AI-generated content and recommendations.

Meanwhile, the technology's power and potential have ensured that progress continues unabated. Ongoing efforts such as AutoGPT and ChatGPT Plugins are pushing the boundaries of AI capabilities even further by integrating internet access and extended memory, allowing the models to carry out autonomous actions in response to user-input goals. These tools hold enormous positive potential, not least for democratizing access to and participation in science, thereby fostering a more inclusive and innovative global scientific community. They have the potential to transform fields such as healthcare, research, law, and education, while promoting cross-cultural collaboration and driving solutions to pressing global challenges. Despite the positive potential, it is essential to address dual-use concerns and potential negative impacts. Scientific freedom under the right to science, we argue, offers a useful lens through which to consider these complexities, balancing the pursuit of knowledge and innovation with the responsible management of risks.

Introduction

This book is about scientific freedom—under responsibility. As protected under international human rights law, science must be used as an instrument for human benefit and must be in harmony with fundamental human rights principles. This means, as the American Association for the Advancement of Science (AAAS) put it in a 2017 statement, that:

> SCIENTIFIC FREEDOM AND SCIENTIFIC RESPONSIBILITY are essential to the advancement of human knowledge for the benefit of all. Scientific freedom is the freedom to engage in scientific inquiry, pursue and apply knowledge, and communicate openly. This freedom is inextricably linked to and must be exercised in accordance with scientific responsibility. Scientific responsibility is the duty to conduct and apply science with integrity, in the interest of humanity, in a spirit of stewardship for the environment, and with respect for human rights.[23]

Many of the major current global challenges—from climate change and pandemics such as COVID-19 to nuclear disarmament and the rising gap between rich and poor—involve science and technology, and digital technologies are ubiquitous in our everyday lives. Science surrounds us to such an extent that it is no wonder many people are afraid of dual-use science research and technology—that is, science and technology that has the potential to be used for great evil as well as for great good.

As we just saw with the example of LLMs, this fear is well-founded and has given rise to demands for citizen involvement in policymaking concerning science as well as for legal regulation. As a result, much current human rights research into science and its role in society and politics focuses on the perspectives of the public, policymakers and citizen scientists. This sometimes leaves out or inadequately represents the outlook of scientists themselves. In this book, we have therefore chosen to look at the issue of scientific freedom primarily, though not exclusively, from the point of view of scholars and scientists. Yet, we will argue, scientific freedom is not primarily concerned with benefiting the scientific community. The protection of this freedom ensures a continuous flow of new ideas and findings, which are essential for societal advancement. The public, therefore, has a vested interest in maintaining scientific freedom, as it is directly linked to the progress and improvement of their own lives. In this sense, scientific freedom becomes a societal obligation, as much as it is a scientific necessity. It is the conduit through which the public can both benefit from and participate in scientific advancement.

Scientific freedom has long been recognized as an essential component for both epistemic progress and the flourishing of democratic societies. Like democracy itself, the argument goes, science thrives in an environment where the exchange of ideas and the pursuit of truth are valued and protected. Short- or even long-term, original hypotheses and experiments may very well lead nowhere. Yet, occasionally, a new paradigm is advanced, or a ground-breaking discovery is made, which is so useful that any number of abortive starts and near misses are tolerated. Sometimes, a single advance outweighs whole generations of scientific dead ends. Indeed, this appears to be how many important discoveries are actually, and have historically been, made.[24]

The freedom to pursue this kind of carefree, blue-skies research is quite remarkable, and is found in few other professions. In what other line of work do employees choose their own projects according to their whim and fancy, and then spend any number of dollars and months or years on it without external oversight and no guarantee of success? Added to this is the difficulty that others outside a given area of research, who do not share an interest in it, will have in understanding not only the methods of a scientific project, but also its aims or relevance. As illustrated by the Golden Fleece Awards, highly important and highly useful science can sound like a bad joke to those who do not understand its implications or its fit within the larger scientific context.[25] Even to those in the know, it is impossible to predict which nooks and crannies of the scientific endeavor may hold the most important insights.

Especially in times of real or perceived economic hardship, it may be quickly forgotten that most kinds of knowledge work synergistically together, that the complexity of today's world requires highly interdisciplinary under-standing, and that the bringing together of either widely differing concepts, or of analogous concepts in differing fields, is one of the primary means of scientific advancement.[26] When the time comes for budget cuts or austerity measures, the argument that the best ideas require diverse inspiration and long incubation times will typically be swept under the rug.

However, despite the evident importance and complexity, discussions on scientific freedom often lack depth and substance. While scientific freedom is readily invoked as a trump card by its defenders, it is just as easily dismissed as empty rhetoric by its opponents. Yet, given the increasing importance of science and technology to everyday life, to the economy, and to the facilitation and protection of individual rights and general well-being, the topic is surely one which deserves more serious treatment. This is all the more so as recent evidence suggests that, despite significant increases in funding, education and numbers of papers and patents, there has been a decline in the rate of scientific innovation over the past several decades.[27] While the causes of this decline are not known, scholars have linked it to changes in scientific funding and management associated with neoliberal economics and politics that began around the time of the observed decline and have continued to influence the conduct of science ever since.[28]

This book seeks to address this gap by providing a comprehensive examination of scientific freedom, with a particular focus on its protection under international human rights law. In doing so, we aim to enrich the ongoing debates on the subject and to offer a fresh perspective by highlighting the added value and relevance of the human rights angle, specifically that of the right to science.

The historical neglect of this right is surprising given the direct relevance of the international human rights law system to science and scientific freedom. Article 15(3) of the International Covenant on Economic, Social and Cultural Rights (ICESCR) explicitly protects scientific freedom by placing obligations on '[the] States Parties to the present Covenant [to] undertake to respect the freedom indispensable for scientific research and creative activity.' Article 15(1)(b) ICESCR furthermore recognizes the right of everyone 'to enjoy the benefits of scientific progress and its applications,' which we can abbreviate to the 'right to science.' As already mentioned, we argue that there can be little enjoyment of the right to science in the absence of scientific progress. Thus, to the extent that scientific freedom is necessary for this progress to happen, it is reinforced and extended by the right to science in Article 15(1)(b). This reading is further supported by Article 15(2) ICESCR, which states that '[t]he steps to be taken by the States Parties to the present Covenant to achieve the full realization of this right shall include those necessary for the conservation, the development and the diffusion of science and culture,' Among these steps may be counted respecting and promoting scientific freedom.

By examining the human rights angle in depth for the first time in this book, we hope to offer several significant contributions to the understanding and protection of scientific freedom. Examining scientific freedom through the lens of international human rights law, we maintain, provides an extra layer of protection for scientific freedom that complements ethical and political arguments, as well as domestic law, thereby helping to solidify its importance. Our analysis reveals that the right to science protects both negative scientific freedom (freedom from interference) and positive scientific freedom (freedom to access and utilize scientific information and resources), encouraging a more holistic understanding of the concept and its implications for researchers and society. Additionally, our examination offers guidance on specific duties and responsibilities that states have, informing policy decisions and fostering greater compliance with international legal obligations.

By emphasizing the protection afforded by international human rights law, we also aim to provide an additional means of fighting unjustified interferences in science and empowering scientists, researchers and advocates to challenge such interferences more effectively. Our exploration of the human rights angle serves, we hope, to further scientific progress by protecting scientific freedom under international law, and by contributing to an environment where the pursuit of knowledge and the exchange of ideas can flourish. Last but not least, we hope that our analysis contributes to a better understanding of the right to science generally, and in particular of the relation between the four different parts of Article 15 ICESCR, their object and purpose, and the extent of the obligations they impose.

Overall, the picture that emerges from our analysis is one of scientific freedom as not only a human right in itself, as outlined in Article 15(3), but also as a necessary and essential, though not sufficient, instrument towards the fulfilment of the right to science in general. According to this understanding, international human rights law places obligations on States Parties to fulfil the right to science, and the way in which States meet these obligations is to ensure scientific freedom. Thus, scientific freedom might be called a constitutive element—a necessary part or constituent—of the right to science. This is a powerful conceptualization, because it brings the many debates concerning scientific freedom into the normatively powerful human rights framework—a framework backed up by seventy-five years of legal tradition.

Treating scientific freedom as a constitutive element of the right to science also has the advantage of bringing a sometimes-nebulous debate within clear boundaries. Human rights law proceeds according to certain rules, with fundamental human rights principles and formal criteria for limitations on the enjoyment of rights. These principles—equality, universality, dignity, non-discrimination, the need to balance the enjoyment of one right against that of other rights as well as weighing competing rights claims—make up a morally weighty and legally binding framework. This may be used to make concrete evaluations of the human rights compliancy of proposed policy moves, with clear reasons to support a judgment.

Our efforts to better understand scientific freedom has led us to the firm conclusion that it is a concept worth taking seriously. To the extent that it really is necessary for scientific progress, few areas of policy can have more important implications for the continued advance of well-being and knowledge. Conversely, to the extent that scientific freedom really can lead to dangerous applications and mistakes, few areas of policy can be more important for protecting the existing health and well-being of our increasingly connected world. Through this comprehensive analysis, we aim to contribute to the ongoing discussion on scientific freedom and the right to science, ultimately promoting a more nuanced understanding of their importance in the modern world.

STRUCTURE AND CHAPTERS

The book is divided into three parts. The first of these, 'Setting the Scene and Providing the Necessary Historical Background,' consists of four chapters. In a brief first chapter, we describe the beginnings of the modern debate on academic and scientific freedom in Germany from the early eighteenth century and onwards, just as we outline key concepts and definitions concerning science and scientific freedom that we use and refer to throughout the present book.

The German debate has been incredibly influential, since from the late eighteenth century, German research universities were perceived as among the best in the world, leading many international scholars to study there. The German model of a research university involved several innovations made by Wilhelm von Humboldt, who based these on ideas derived from the philosophy of Friedrich Schleiermacher. These included the equal standing of students and professors in seminars, such that they could work together as graduate students and their supervisors now commonly do; the freedom of professors to teach as they saw fit, and the corresponding freedom of students to learn as and what they please; and finally, the idea of *Freiheit der Wissenschaft*, or intellectual freedom. Especially American scholars were apt to take these ideas back to their home country as they returned from research or study trips.

The second chapter of Part I presents the method used and the results obtained of a systematic review of the scholarly literature on scientific freedom—a review which provides the necessary context and background against which to assess the human rights dimension of scientific freedom. Based on these results, this second chapter also outlines the three major justifications for scientific freedom in the included studies. In addition to its necessity in the pursuit of truth, we found two other major justifications for scientific freedom in these studies. The first is political and points out that scientific evidence is crucial for correctly understanding many politically sensitive issues, such as climate change and welfare policies. If there were quality information about the likely effects of various welfare policies, this argument goes, people could weigh this information against their personal feelings on the matter, but at least these feelings would be scientifically informed. To achieve this, science must be free from political control; otherwise, there will always be an incentive to mess with the results in a way which favors a certain political angle.

The second justification, as previously alluded to, was that scientific freedom is required by international human rights law. This approach to scientific freedom is intellectually interesting, not only because of its promise and previous neglect, but also because it subsumes, to a significant extent, the other two arguments. The epistemic argument that scientific freedom is necessary for scientific progress has substantial implications for State obligations under the right to science. Yet the political dimensions, according to which access to objective information is necessary for democracy, can also be seen as an integral aspect of the right to science.

To provide the context necessary to substantiate these claims, the last two chapters of this first part of the book offer an in-depth examination of the right to science and scientific freedom, respectively. The chapter dedicated to the right to science, 'Taking Human Rights to the Next Level: The Right to Science, History and Content,' explores its antecedents, its history and drafting, as well as subsequent scholarship and soft law instruments relating to the right. First mentioned in Franklin Delano Roosevelt's 1941 State of the Union address, the right to enjoy the benefits of scientific progress has its origins in the social and political climate of the 1920s and the 1930s. With the rise of totalitarian regimes and their suppression of basic human rights, the importance of scientific progress and economic security in maintaining peace and stability became more evident. This insight led to the inclusion of the right to science in the first international human rights document, the American Declaration of the Rights and Duties of Man of 1945, which fed into the drafting process of the Universal Declaration of Human Rights (UDHR) of 1948 and, in turn, the International Covenant on Economic, Social and Cultural Rights (ICESCR) of 1966.

The drafters of these documents highlighted key themes related to the right to science, including access to and participation in science, scientific freedom, and the balance between intellectual property protection and the role of science in society. The drafting history of the ICESCR reveals the importance that delegates placed on scientific freedom and their belief that it was essential for scientific progress. Article 15(3) ICESCR, heavily influenced by the UDHR and the UNESCO constitution, enshrines scientific freedom as a negative freedom, ensuring that individuals are not prevented from engaging in scientific activities.

Subsequent soft law instruments like the 2009 Venice Statement on the right to science, the 2012 report of the Special Rapporteur in the Field of Cultural Rights, the 2020 General Comment on Science and Economic, Social, and Cultural Rights, as well as academic scholarship, support the interpretation of scientific freedom as a constitutive element of the right to science. The General Comment and the Special Rapporteur's report outline state obligations to respect, protect, and fulfill the right to scientific freedom. They highlight the significance of both negative and positive obligations, including the protection of researchers from undue influence, promoting academic and scientific freedom, ensuring equal access and participation, and fostering an enabling environment for the conservation, development, and diffusion of science.

While scientific freedom is integral to the right to science, it is not absolute and may be subject to limitations under clearly defined criteria as laid out in Article 4 ICESCR. This recognition of potential limitations balances the need for scientific freedom with competing demands on resources as well as the legitimate concerns of states in promoting the general welfare of society in ways other than advancing science. Understanding the importance of scientific freedom as a constitutive element of the right to science enriches our interpretation of this right under international human rights law, we argue in Chapter Three; it also emphasizes the crucial role that scientific freedom plays in promoting scientific progress and enhancing the overall well-being of individuals and societies.

Chapter Four then looks at scientific freedom and how it relates to Article 15(1)(b), 15(2) and 15(4) ICESCR. For the right to science to work as intended, scientific freedom cannot stand alone. When States become parties to international human rights treaties, they assume responsibilities under international law to respect, protect, and fulfil human rights, and to enact the necessary domestic measures and legislation.[29] With regard to the right to science, this means that, in addition to respecting scientific freedom, States Parties to the ICESCR are obligated, according to Article 15(2), to take steps towards conserving, developing, and diffusing science. They must also recognize the importance of international contacts and co-operation in the scientific field, as outlined in Article 15(4). The major focus of the chapter is on the dissemination part of Article 15(2) as it relates to scientific freedom and overlaps with the global aspect of Article 15(4).

The second part of the book, 'The SAFIRES Model,' consists of two chapters. These chapters aim to integrate the lessons of the previous chapters into a coherent understanding of scientific freedom as an integral element of the right to science. In an effort to make this right more actionable and more easily useable, in the first chapter of this second part of the book, we develop and describe the SAFIRES (Scientific and Academic Freedom as an Integral element of the Right to Enjoy the benefits of Science and its applications) model. This model, which is based on and extends our previously published Four Step test for the evaluation of scientific policies and their compliance or otherwise with the right to science, is meant to facilitate the practical application of right to science-based scientific freedom standards to policymaking, evaluation and analysis.[30] Our hope is that those tasked with understanding and implementing human rights obligations, as well as those laboring to hold governments to account for these, will find the SAFIRES model a useful tool in litigating, advocating, legislating and otherwise affecting science-related policy.

The next chapter applies the SAFIRES model to three international instruments of great relevance to scientific freedom as well as science and society more broadly. These are the 2017 UNESCO Recommendation on Science and Scientific Researchers; the 2020 General Comment; and the 2021 UNESCO Recommendation on Open Science, respectively. In applying the model to these instruments, our aim is both to provide practical examples of the model's usefulness (and by extension of the understanding of scientific freedom advanced here), and to draw attention to and critique those provisions in these instruments which appear to us suspect from the point of view of scientific freedom as a constitutive element of the right to science.

In the third and final part of the book, we yield the floor to four colleagues whom we have invited to share with us, and the readers of this book, their experiences with working with the right to science in practice. First, Mikel Mancisidor, former member of the Committee on Economic, Social and Cultural Rights, CESCR, and main author of CESCR General Comment No. 25 on the right to science, talks about the creation of this very important General Comment. Next, Cesare Romano, Professor of law at Loyola Law School in Los Angeles, writes about preparing and submitting to the CESCR, as a part of the 'individual communications' process of the ICESCR, a case on the violation of the right to science. And finally, Malene Nielsen and Carsten Staur, Deputy, Permanent Delegation of Denmark to UNESCO, and Danish Ambassador to UNESCO, respectively, share with us their thoughts on and their ongoing work within UNESCO on 'Defending science, knowledge, and facts: The UN and Scientific freedom of expression.'

PART I

Setting the Scene and Providing the Necessary Historical Background

Chapter 1

Historical Background and Key Concepts and Definitions

Arguments for the freedom of speech and press are ancient, as the following quote from Euripides, who lived ca. 480–406 B.C., illustrates:

> *This is true Liberty when free born men*
> *Having to advise the public may speak free,*
> *Which he who can, and will, deserv's high praise,*
> *Who neither can nor will, may hold his peace;*
> *What can be juster in a State than this?*

—Euripides, *The Supplicants*

However, modern notions of scientific and academic freedom stem from the 18th century. In this opening chapter, we provide the historical background for these modern notions and the way in which they were justified. Then, in the second part of the chapter, we outline a few concepts and definitions that are key to understanding scientific freedom and what it entails. In addition to 'scientific freedom' itself, these key concepts are 'science,' 'freedom from interference in science,' 'freedom to participate in science,' 'freedom to access science,' and 'scientific, academic, and related freedoms.' We end by proposing, for the purposes of this book, our own definition of scientific freedom.

HISTORICAL BACKGROUND

Scientific freedom has historically been closely related to the idea of academic freedom. The latter concept can be traced back at least to the early Enlightenment, with philosophers such as Nicholas Gundling arguing as early as 1722, in a speech given at the University of Halle, that:

Step by step must the summit of truth be scaled . . . it is virtually impossible for even the most diligent not to slip here and there . . . before they reach the place where there is no more occasion for erring and slipping. But now imagine that the erring were not tolerated . . . Who could still ascend to that highest pinnacle of truth? Therefore, freedom must be conceded to reason.[1]

Implicit in this early formulation is a justification of academic freedom: without it, those who seek knowledge cannot make the mistakes necessary for discovery. Another early epistemic argument for academic freedom was given by the philosopher Christian Wolff in 1728:

One person acknowledges the truth taught by the other and uses it to detect further truth. Another one points out an error that has been made or improves upon it; and the one who has made it acknowledges it and tries to correct it, if it has not already been corrected by others. Thus is the growth of the sciences advanced by means of joined forces.[2]

What Wolff described here is that innovative ideas do not occur in a vacuum. The scholarly process is iterative and collective, each scholar standing, to adapt Newton's famous phrase, "on the shoulders of others."[3] Scientific progress relies on the ability of scholars to express their ideas freely and to engage in critical analysis of the work of others, correcting mistakes and advancing understanding through open and rigorous inquiry.

During the eighteenth century, academic freedom became a prominent topic at German universities, particularly at the Universities of Halle, Jena, and Göttingen. The latter university was founded early in the eighteenth century yet already contained within its statutes a provision guaranteeing '*honesta quadam docendi sentiendique libertas* [the responsible freedom of teaching and thought].' On the campuses of Halle and Jena, a series of prominent philosophers discussed the necessity of defending and the best ways of understanding academic freedom.

These discussions set the background for the development, beginning in the early nineteenth century, of the German research university. Wilhelm von Humboldt founded the University of Berlin in 1810. He initiated significant modifications in both secondary and tertiary education systems. Two major changes were pivotal to the emergence of the contemporary research university. The first was the focus on research as opposed to instruction. The second, closely connected to the first, was a transformation in the roles of students from passive information receivers to active participants in seminars, laboratories and research activities.

These features were partly enshrined in the German conception of academic freedom, which at this point consisted of three parts: *Lehrfreiheit* (freedom of teaching), *Lernfreiheit* (freedom of learning) and *Freiheit der*

Wissenschaft (scientific freedom).[4] As we describe below, in the section on 'Key Concepts and Definitions,' the German word *Wissenschaft*, though often translated into English as *science*, is a broader term that also includes the social sciences, humanities and, to some extent, the arts. In modern English, the term 'science' is commonly associated with natural and social sciences. This contemporary English usage has only become widespread since the late nineteenth century, however, and earlier interpretations resembled a broader concept of intellectual study, similar to the German term '*Wissenschaft*.'[5]

The German conception of academic freedom, and the system it helped underpin, stood in stark contrast to the collegiate universities of America and other academic systems. Because of this, they helped attract many foreign scholars to the seminars, public lectures and laboratories of German universities. Notably, several generations of American scholars attended German universities during this time. When they returned home, they brought with them the concepts of academic and scientific freedom. These were much discussed in American writing at the time.[6]

However, the three underpinning concepts had specific connotations in Germany which were not easily transplanted across the Atlantic. *Lehrfreiheit* referred to the freedom of full and associate professors, who in Germany were classified as civil servants, to operate outside some of the laws and regulations which affected civil servants more generally. In the United States, professors were typically considered to be employees of the board of governors of the university. *Lernfreiheit* was understood to mean that the universities disavowed any authority over students beyond that of administering exams and granting degrees; whereas, in the United States, colleges typically retained *in loco parentis* responsibilities over their charges. Finally, an important component of *Freiheit der Wissenschaft* was understood to be academic self-government: the independence of the university as an institution from interference by political or other outside powers. In the United States, the university as an institution was dependent on the whims of the members of the governing body, who were typically donors or political actors.[7]

Thus, national differences forced American scholars who had studied in Germany to adapt the concepts behind academic freedom to their native context. The politically motivated firing of an economics professor, Edward Ross, from Stanford in 1900, and the subsequent firing of two professors and two instructors from the University of Utah in 1915 lent urgency to the issue. In response, a team of professors led by Arthur Lovejoy and John Dewey established the Association of American University Professors (AAUP) in 1915. One of their first achievements was the writing of a report, known as the 1915 *Statement of Principles*, that served to crystallise the American interpretation of the German concepts related to scientific and academic freedom.[8] This report focused heavily on the independence of faculty from administrative and employer pressure, as well as the freedom of extramural speech, arguably to the exclusion of other important issues.

The 1915 report was eventually supplemented by a set of guidelines set forth in a 1940 document co-authored jointly by the AAUP and the Association of American Colleges, which represented university management. This document concentrated on three principles, again to the exclusion of others: firstly, the freedom of teaching (with a single exception: an injunction against bringing irrelevant controversy into the classroom); secondly, extramural speech; and thirdly, freedom of inquiry (which could legitimately be restricted in case it interfered with the other functions of professors, such as teaching, and in case any external monies were derived from the research unless this was based on an understanding with the administration). This statement remains, to this day, among the most important documents for the interpretation of the American conception of academic and scientific freedom.[9]

The 1940s and the couple of decades that followed it saw great developments in the legal, human rights and scholarly fields related to scientific freedom. Perhaps most importantly, the United Nations' 1948 Universal Declaration on Human Rights (UDHR) contains several articles relevant to academic and scientific freedom. These include:

- Articles 9–12, which deal with arbitrary detention, the right to a fair tribunal, the rights to be presumed innocent and not to be punished for an act that is not an offense, and the right against arbitrary interference with privacy.
- Article 13, the right to freedom of movement.
- Article 14(1), the right to seek asylum.
- Article 18–20, the rights to freedom of thought and conscience, opinion and expression, association and peaceful assembly.
- Article 21(1–2), the rights to seek office and to have equal access to public service.
- Article 22, the right to social security and the full realisation of one's dignity and personality.
- Article 23(1–4), the rights to work, equal pay, fair remuneration, and to establish or join a trade union.
- Article 24, the right to rest and leisure.
- Article 25(1–2), the rights to an adequate living and to special assistance for motherhood and children.
- Article 26(1–3), the right to education.
- Article 27(1–2), the rights to culture, arts and science, and the right to protection of the moral and material interests resulting from scientific, literary or artistic productions.

Although all of these are important, a case can be made that it is Articles 26 and 27, which concern education, culture and science, Articles 18 and 19, which concern freedom of thought and expression, as well as Article 23 on the right to work that are the most directly relevant to scientific freedom.

The moral and political commitments expressed in the Universal Declaration were turned into law by the 1966 International Covenant on Economic, Social and Cultural Rights (ICESCR). The ICESCR came into effect in 1976 and includes analogues of most UDHR rights, including the ones we consider most relevant to scientific freedom:

- Articles 6–8, the rights to work, fair remuneration, equal pay, rest and leisure, a decent living, safe and healthy working conditions, to strike, and to form or join a trade union.
- Article 13(1–2), the right to education.
- Article 15(1–4), the rights to science, culture, protection of moral and material interests resulting from literary, artistic or scientific works, freedom of scientific research and creative activity, and international cooperation in culture and science.

The United States, having signed but not ratified the ICESCR and disputing the status of the UDHR as international customary law, is still bound by the law of treaties to refrain from any action that would defeat the object or purpose of the ICESCR.[10] Nevertheless, discussion of academic and scientific freedom in the US context shows considerably less emphasis on human rights, leaning instead toward scholarship and legal, especially constitutional, analysis, as well as legal and professional protection of scholars.

Unlike the constitutions of several other countries, including Germany (Article 5), Italy (Article 33), Slovenia (Article 59) and Greece (Article 16), neither academic nor scientific freedom are enshrined in the constitutions of the United States or Canada.[11] This situation has led to extensive discussion concerning whether these freedoms are derivate from or contained in the First Amendment to the US constitution, which concerns free speech. Scholars in this domain largely appear to agree that there is some constitutional protection for scientific speech under the First Amendment.[12]

The case for constitutional protection of academic freedom relies on deriving such protection from other clauses. Prime among these is the First Amendment, which protects the freedom of expression and speech. This can be seen at work in the Supreme Court's judgment in *Keyishian v Board of Regents*.[13] The background to that case was a New York State law which prohibited the hiring or continued employment of state employees who were members of any organisations that advocated overthrowing the US government or were found treasonous or seditious. In addition, the law required state employees, including professors at public universities, to sign a loyalty oath. Several instructors, among them Keyishian, were terminated under this law and appealed their termination to the Supreme Court.

The Court, in a 5–4 decision, overturned the New York State law on the grounds that it was vague and overbroad and thus liable for misuse. However, the case is best remembered for its endorsement of academic freedom in Brennan J.'s judgment:

> Our Nation is deeply committed to safeguarding academic freedom, which is of transcendent value to all of us and not merely to the teachers concerned. That freedom is therefore a special concern of the First Amendment, which does not tolerate laws that cast a pall of orthodoxy over the classroom. 'The vigilant protection of constitutional freedoms is nowhere more vital than in the community of American schools.' The classroom is peculiarly the 'marketplace of ideas.' The Nation's future depends upon leaders trained through wide exposure to that robust exchange of ideas which discovers truth 'out of a multitude of tongues, [rather] than through any kind of authoritative selection.'

Brennan J.'s judgment illustrates the view that academic freedom is a "special concern" of the First Amendment. There are scholars, however, who would point our attention to the important distinction between academic and scientific speech, plausibly protected by the First Amendment, and scientific conduct, where protection is less plausible.[14] Thus, aspects of scientific work—e.g., accessing and reading the reports of other scientists or publishing your own—are plausibly interpreted as having some First Amendment protection, whereas others—e.g., genetic engineering of viral strains in an attempt to aerosolise them—are less, if at all, plausibly interpreted as having anything to do with expression and speech.

In the aftermath of the Second World War, rights bearing on scientific freedom began to be recognised in supranational and regional agreements. The EU Charter, for example, recognises the rights to freedom of thought and conscience (Article 10[1]), freedom of opinion and expression (11[1]), freedom of assembly and association (12[1]), freedom of the arts and sciences (13), the right to education (14[1]), the rights to work (15) and conduct a business (16), and the right to property (17), which includes the right to intellectual property protection (17[2]).

The Banjul Charter (African Charter on Human and People's Rights) recognises the rights to freedom of conscience and profession (Article 8), the right to receive (9[1]) and disseminate or express information (9[2]), freedom of assembly (10[1]) and movement (11), the right to fair work (15), the right to education (17[1]) and culture (17[2]).

The American Human Rights Charter recognises the rights to freedom of conscience (Article 12), thought and expression (13), assembly (15), association (16), and movement (22).

The Association of Southeast Asian Nations (ASEAN) Human Rights Declaration recognises the rights to freedom of movement (15), thought (22), opinion and expression (23), work (27), adequate standard of living (28), education (31) and to freely take part in cultural life, to enjoy the arts and the benefits of scientific progress and its applications (32).

Finally, the Arab Charter on Human Rights recognises the rights to freedom of assembly and association (24), movement (26), thought (30), opinion and expression, as well as to seek, receive and impart information (32), to work (34), an adequate standard of living (38), education (41) and to take part in cultural life and to enjoy the benefits of scientific progress and its applications (42). The latter right is complemented by an undertaking of signatory States to 'respect the freedom of scientific research and creative activity.'

Apart from nations and regional bodies, international organisations have also worked to interpret and fulfil rights relating to scientific freedom. UNESCO's unique role and remit, being officially responsible for three areas—education, science and culture—directly related to scientific freedom, has qualified it to undertake a number of initiatives culminating in recommendations and instruments which bear on the freedom of scientific inquiry. These include the Recommendations on the Status of Teaching Personnel (1966 and 1997), the Convention Against Discrimination in Education (1960), the Recommendation on the Recognition of Studies and Qualifications in Higher Education (1993), the Recommendation on the Status of the Artists (1980), the Convention on Diversity of Cultural Expressions (2005) and, perhaps most importantly, the Recommendation on Science and Scientific Researchers (2017), and the Recommendation on Open Science (2021). These instruments contain several recommendations and regulations (in the case of Conventions) that are relevant to current issues in scientific freedom.

KEY CONCEPTS AND DEFINITIONS

Throughout the present book, we use and refer to certain key concepts and definitions. Among these are 'science,' 'scientific freedom,' 'freedom from interference in science,' 'freedom to participate in science,' 'freedom to access science,' and 'scientific, academic, and related freedoms.' They will be briefly outlined below.

Science:

In modern English, the term 'science' is commonly associated with natural and social sciences, but its definition is more complex than that. Merriam-Webster offers several interpretations, including 1) 'knowledge or a system of knowledge covering general truths or the operation of general laws especially as obtained and tested through scientific method,' 2) 'a department of systematized knowledge as an object of study,' and 3) 'the state of knowing: knowledge as distinguished from ignorance or misunderstanding.'[15] However, these contemporary English usages have only become widespread since the late nineteenth century, and earlier interpretations resembled a broader concept of intellectual study, similar to the German term *Wissenschaft*.[16] The German Constitutional Court defines *Wissenschaft* as any activity which 'in content and form is to be seen as a serious, structured attempt at reaching the truth.'[17] As emphasised by the Constitutional Court, the important thing is that a sincere, *bona fide* attempt is made at producing new knowledge, rather than merely advancing a particular opinion.

The broader concept of science is not unique to German. Other languages, such as the Nordic languages and Dutch, have terms that share etymological and conceptual similarities with *Wissenschaft*. This distinction is crucial in international human rights law, where no single tradition can be assumed to represent the general understanding of all parties to a treaty like the ICESCR. In fact, Article 33 of the Vienna Convention on the Law of Treaties (VCLT) stipulates that when a treaty is authenticated in multiple languages, each version is equally authoritative, and any divergence in meaning should be reconciled with an interpretation that best aligns with the treaty's objective and purpose, unless a prevailing text is specified. Comparing the authentic language versions of the UDHR and ICESCR reveals that the terms used in Russian, Arabic, French, Spanish and Chinese texts all possess a broader scope than the modern English term for 'science.'[18]

For these reasons, we here adopt the broader sense of 'science,' including both natural and social sciences but also the humanities.

Scientific Freedom

As we saw above, the modern concept of scientific freedom can be traced back to the German concept of *Freiheit der Wissenschaft*. The term originally referred to academic self-government, that is, the independence of universities from administrative and other impositions from the secular and religious authorities.

Many contemporary commentators include in their definitions of scientific freedom this element of freedom from outside interference. Thus,

Wilholt states that scientific freedom 'can be understood just to imply that the scientists involved in research should themselves decide which projects and approaches to pursue,' an element which he calls the 'freedom of ends.'[19] However, modern usage is often broader and reflects in addition the 'freedom of means,' that is, 'the claim that society or the state must provide for the resources required to conduct all the research that the scientists deem important.'[20] The terms 'negative' (referring to the absence of interference) and 'positive' (referring to the presence of required resources) freedom are often used to make this same distinction. Holm explains:

> Scientific freedom understood as a negative freedom or a liberty is simply the freedom of an individual researcher or a group of researchers to pursue any scientific question or research program without outside interference. Scientific freedom understood as a positive freedom is more difficult to define succinctly, but includes the claim that society should materially and in other ways support science. Society should remove obstacles to science and it should, within the general envelope of its resources also support valid scientific endeavours.[21]

In the human rights context, rights and obligations are analysed according to an even more detailed classification scheme known as the tripartite typology, which classifies State obligations under the three headings of 'respecting,' 'protecting' and 'fulfilling' a right. To respect a right means that a State party must not itself infringe it; protecting it goes beyond this in obliging State parties to prevent third parties from violating the right; and, most demandingly, fulfilling a right means bringing about the conditions necessary for its full enjoyment. The latter two types of obligations can be analysed as containing positive obligations, while the first contains negative obligations.

For the purposes of this book, we can be even more specific. For reasons relating to the history, object and purpose, and current understanding of the right to science, all of which are explored in much more detail in following chapters, we define scientific freedom as consisting of the following three elements:

- Freedom from interference in science
- Freedom to participate in science
- Freedom to access science

Freedom from Interference in Science

This element largely corresponds to negative scientific freedom, or freedom from external interference in science. It implies that key decisions relating to the scientific process, such as the choice of research topic and method(s), how to evaluate results, and whether to publish them, should be taken by scientists themselves. Moreover, it implies that these decisions should be taken without influence from politicians, the public, or other interest groups.

This freedom can never be absolute. Like everyone else, scientists are subject to laws meant to protect the health and interests of others. In addition, there are norms deriving from the scientific community as to how science should be conducted responsibly. There are also guidelines specific to certain areas of science, such as those governing ethical experimentation in human subjects research, for example, in medicine. Professional scientists can also be subject to institutional and contractual obligations as a condition of employment. Not all these restrictions are benign or appropriate, but many result from long experience and debate and are accepted as uncontroversial.

For the purposes of this book, we define this element of scientific freedom as the *absence of limitations on the choice of scientific research topics and methods as well as on the interpretation and dissemination of results that do not meet the Article 4 ICESCR criteria for legitimacy.* These criteria will be explored in more detail in later chapters.[22]

Freedom to Participate in Science

Freedom from external interference is necessary but not sufficient for doing science. In addition, appropriate conditions for participating in science must be available. This means, at a minimum, that no one is discriminated against when trying to conduct science. It also means that the education, equipment, and basic resources necessary for carrying out research projects are available.

Apart from its non-discrimination component, this aspect of scientific freedom corresponds more closely to positive than to negative freedom. Fulfilling this aspect of scientific freedom is also likely to be harder and more expensive than respecting and protecting the freedom from interference. For this reason, and because scientific freedom has historically been most frequently analysed according to its negative freedom (or liberty) dimension, the notion of scientific freedom requiring positive actions and extensive State expenditure may seem surprising and perhaps even controversial.

Without the provision of these basic ingredients of the scientific enterprise, however, science would be dependent on private investment. This means that much less scientific progress would be made. Without scientific progress, the right of everyone to benefit from that progress and its applications cannot be fulfilled. For this reason, we argue here that Article 15(1)(b)ICESCR requires the positive provision by States of the education, equipment, and other resources required to conduct science. This obligation is reinforced by Article 15(2) ICESCR, which holds that 'The steps to be taken by the States Parties to the present Covenant to achieve the full realization of this right shall include those necessary for the conservation, the development and the diffusion of science and culture.'

Freedom to Access Science

A further element required for the freedom of science is the freedom to access scientific information, data, and hypotheses. These may be stored in databases, journal articles, books, on preprint servers, or in other locations, many of which are not openly accessible by default but require paid individual or institutional subscriptions. The higher the barriers to accessing the intangible ingredients of scientific progress, the less such progress there will be.

Freedom to access science extends beyond data and information to the tangible realm of scientific instruments, equipment and so on. As noted above under the freedom to participate in science, many of these resources are scarce and expensive and thus may require public funding and investment. However, others are inaccessible altogether because they are protected by intellectual property rights. To the extent that these inhibit the more widespread use of scientific tools and technologies necessary for advancing science, they are problematic from the point of view of scientific freedom under the right to science.

Scientific, Academic, and Related Freedoms

It is important to differentiate scientific freedom from related freedoms, as the level and extent of protection may vary, and these associated freedoms are protected by other human rights, laws and arguments. Although scientific freedom as a human right applies universally to every human and to every branch of research, it does not imply that each of the three elements of scientific freedom applies equally to professional scientists and amateurs. There may be legitimate reasons for restricting access to dangerous experimental substances for amateurs, students and citizen scientists, for example. Nevertheless, all individuals conducting science should enjoy as much freedom of access to scientific information and knowledge as possible and as compatible with the limitations criteria in Article 4 ICESCR.

A distinct and higher level of freedom, known as academic freedom, is granted to scientists researching in publicly funded universities and government research organisations. One argument for recognising a higher degree of intellectual freedom rights for academics stems from the public good orientation of their research. Since university researchers work for the public benefit rather than private gain, the absence of the profit motive generates a presumption of greater public benefit. Another justification for academic freedom, explored by Beiter and Barendt, is the unique "role of universities as venues for the free exchange of ideas."[23] Unlike private research institutes, universities do not set the research agenda and are not driven by ideological or market-based interests. This special status of universities justifies the recognition of the related protection of academic freedom. While every human enjoys scientific freedom, only academics—those who carry out research in universities and publicly funded research institutes—enjoy academic freedom.

Scientific freedom is also related to, but distinguishable from, freedoms of opinion and expression. Although aspects of the right to science are protected by these freedoms in some legal systems which do not recognise broader, science-related rights and freedoms, freedoms of opinion and expression carry narrower protections in the context of science, because they do not extend to conduct. Moreover, as explained by Beiter, while freedom of expression and opinion has been understood to extend even to known falsehoods, this is not the case for scientific and academic freedom, since these are based on the different rationale of promoting knowledge and seeking truth.[24]

Based on the above, for the purposes of this book we define scientific freedom as follows: *Scientific freedom is the right to pursue scientific research without undue external influence, to participate in scientific activities, and to access scientific knowledge and resources. This includes freedom from interference (negative freedom) and the availability of resources and support (positive freedom) and extends to all branches of learning, not just the natural or social sciences, provided that a sincere, bona fide attempt is made at advancing knowledge, rather than merely promoting a particular opinion.*

Chapter 2

A Systematic Review of the Scholarly Literature on Scientific Freedom: Methods and Results

In this chapter, we offer the results of a systematic review of scholarly literature on scientific freedom. This review not only provides the necessary context and background against which to assess the human rights dimension; it also backdrops the rest of the chapters of this book.

The preparation for the writing of this book included extensive consultation of scholarly books and articles dedicated to, or containing a treatment of, the concept of scientific freedom. In locating these sources, we adopted a technique commonly used in the medical sciences for the synthesis of evidence relating to a specific topic. Known as a *systematic review*, this methodology sets out, in advance, the search strategy to be used in finding source material, as well as criteria used to determine whether these works are relevant to the topic and should thus be included in the subsequent analysis.

The motivation behind carrying out a systematic review was to provide an overview of the English-language literature on scientific freedom which, if not exhaustive, was at least representative enough for us to generalise and make inferences about the scholarship on the topic. Using this approach, we located a total of eighty-four articles and books matching our inclusion criteria. Next, we read these articles and books and identified twenty distinct topics within the included studies. These ranged from specific issues mentioned by few studies (such as whether Institutional Review Board [IRB] oversight constitutes a breach of scientific freedom) to broad, conceptual questions (such as the relation of scientific freedom to policymaking and democracy) discussed by many. Indexing each study according to whether it raised or touched upon each of these twenty topics allowed us to identify the key features and issues of the scholarly debate on scientific freedom as measured by the frequency and depth of their discussion.

In addition, we grouped some of these features into three thematic clusters based on conceptual similarity: 'Understanding,' 'implementing' and 'tensions.' Each of these clusters encompasses several thematic features; the clusters 'understanding' and 'implementing' each include seven features, while the cluster 'tensions' includes only six features.

Schematically, our three clusters and their thematic features look like this:

Understanding	Implementing	Tensions
Justifications/grounding	Policy/Democracy	Access
Interpretation	Government intervention	Inequality
History	Science v Society	Human rights
Connection to free speech/press	Restrictions	IRBs
Conflicts	Responsibilities	Incentives/funding
Application	Ethics	Dual Use
Reasoning	Legal	

We adopted this categorisation for ease of analysis and presentation. However, it should be noted that there was some overlap between clusters: for example, the ethical and legal focus of many articles was relevant not only to 'implementing,' but also to 'understanding' scientific freedom and to several of the specific 'tensions' highlighted in the literature. Likewise, the 'interpretation' and 'justifications/groundings' features both involve attempts at clarifying the normative and practical implications of scientific freedom based on a better understanding of the concept itself. Similarly, the 'restrictions,' 'conflicts' and 'government intervention' features all relate to the legitimacy or otherwise of state intervention into scientific freedom, while 'IRBs,' 'open access,' 'inequality/discrimination,' 'human rights' and 'incentives/funding' all address specific ways in which scientific freedom may conflict with state priorities or regulation.

Given this overlap between clusters and their thematic features, in the following presentation of the literature, we keep the two clusters 'understanding' scientific freedom and 'implementing' scientific freedom (and use them as sub-headings for the two parts of this chapter), but we fold the third cluster, 'tensions,' and its six thematic features into our exposition of the other two clusters.

Among the issues we focus on in the first part of the chapter, 'Understanding Scientific Freedom,' are the three major justifications for scientific freedom in the included studies. In the second part of the chapter, 'Implementing Scientific Freedom,' the topic of 'dual use' and the regulation of science, whether by the government or by scientists themselves, plays a crucial role.

UNDERSTANDING SCIENTIFIC FREEDOM

This cluster includes seven features related to the conceptual and practical understanding of scientific freedom. Given the importance and relevance of historical perspectives to this topic, we dedicated a separate section in Chapter One to the conceptual history of scientific freedom. As indicated in that section, the German concept of scientific freedom, from which much of our current understanding is derived, was originally contained within the larger concept of academic freedom.

Several articles made clear that scientific freedom can and should be contrasted to the related concepts of academic freedom, intellectual freedom and freedom of speech, each of these terms being protected in different ways and covering distinct, though related, concepts. As we stated in the 'Key Concepts and Definitions' section in the previous chapter, we understand 'scientific freedom' to be the right to pursue scientific research without undue external influence, to participate in scientific activities, and to access scientific knowledge and resources. This includes freedom from interference (negative freedom) as well as the availability of resources and support (positive freedom), and extends to all branches of learning, not just the natural or social sciences. While scientific freedom, as a universal human right, applies equally to every person regardless of education and profession, there may nevertheless be compelling reasons to consider education and professional status in decisions concerning allocation of scarce resources.

Academic freedom, however, is a distinct and elevated level of freedom granted to researchers in universities and government research organisations. The public good orientation of academic research and the unique role of universities as forums for the free exchange of ideas justify this distinction. Considering the massive resources and manpower modern tech companies have, the conditions of freedom for private sector companies are also important. However, researchers in private companies have often contracted to carry out specific research. Private companies are also not bound to implement human rights standards in the same way that governments, including public universities, are. While scientific freedom is related to freedoms of opinion and expression, it extends beyond speech to aspects of scientific conduct.

Academic and scientific freedom are also protected on the professional level. As mentioned in the historical section in Chapter One, the American Association of University Professors (AAUP)'s 1940 guidelines provide much of the basis for the understanding of academic freedom in the US. Described by one scholar as 'very soft law,'[1] these guidelines focus prominently on issues relating to intra- and extramural speech as well as freedom of teaching and research; they do not have separate sections dealing with science and scientific freedom. In practice, the professional dimension to academic freedom afforded by the AAUP is nearly only invoked in the context of faculty tenure decisions.[2]

The situation in the United States contrasts to that in the European Union. Not only do the constitutions of seventeen of the twenty-seven EU member-states explicitly protect scientific freedom; Article 13 of the Charter of Fundamental Rights of the European Union of 2007 also directly refers to both academic and scientific freedom.[3]

'JUSTIFICATIONS/GROUNDINGS' AND 'CONNECTIONS TO FREEDOM OF SPEECH/PRESS'

Scientific freedom grants freedoms to scientists which are not enjoyed by other individuals and professionals. These greater freedoms require justifications. We found several in the included literature.

The Epistemological Argument

The most important and most frequently mentioned justification for scientific freedom is the argument that it is necessary for scientific progress. Thus, the poet Milton, in his defense of free publishing before the Houses of Lords and Commons, thought it 'not only by great authorities brought together, but by exquisite reasons and theorems almost mathematically demonstrative, that all opinions, yea errors, known, read, and collated, are of main service & assistance toward the speedy attainment of what is truest.'[4]

Many of the included studies relied heavily on this justification. Some considered it nearly self-evident: 'The crucial rationale for the protection of freedom of science must be seen to lie in the fact that it makes possible the discovery of the truth.'[5] Others pointed to historical examples of major scientific breakthroughs which were the result of curiosity-driven research and tinkering with new techniques and methods. Among the very many examples mentioned are: Discovery of the positron; the mathematical foundation for understanding electromagnetism; NMR spectroscopy, on which MRI imaging is based; X-rays; nuclear fission; antibiotics, antibodies, immunosuppressive drugs, the structure of DNA; echocardiography, cardiopulmonary bypass and the effect of adrenaline on blood pressure.[6] It should be noted that these and other examples in the included literature predominantly concern the natural and physical sciences, rather than the social sciences or humanities. This is in keeping with the narrower usage of 'science' in English noted previously.

One of the main arguments in support of the epistemological argument is that due to the inherent fallibility in predicting the success of research projects, scientists should have the freedom to choose various research paths, as this diversity could lead to unexpected groundbreaking discoveries. Wilholt states the argument as follows:

> [all] prior judgments about the fruitfulness of research projects are fallible. It cannot be precluded that projects which are at present not recommendable according to widespread standards will turn out to be groundbreaking. Therefore, scientists should choose their approaches and projects freely, such that a wide variety of approaches ends up being pursued. Some of these will prevail and lead to new knowledge, but it is impossible at any time to predict which ones these will be.[7]

Wilholt goes on to note that for the epistemological argument to succeed, it must only demonstrate a relative advantage over other means of scientific organisation in the diversity of research topics explored, under the assumption that the diversity of approaches taken when scientists enjoy freedom is the driving mechanism behind scientific development. When scientists are free to explore various methods of addressing a problem, it is likely that they will in fact make use of various methods in the face of competition to derive the benefits of being the first to solve a scientific problem.[8]

Given that there are so many examples of scientific advances being made only after orthodox means have been exhausted and unorthodox methods embraced, it would appear prudent to allow for as wide a variety of scientific approaches as possible. Scientific freedom is arguably the best way to achieve this, given the difficulty that other possible models—e.g., centralised control over research topics—face in obtaining the detailed knowledge of the skills, ideas and interests of individual scientists and their ideal configuration into teams. Thus, decentralisation at the level of the individual is plausibly thought to be more efficient than centralisation at the level of national or regional science planning. Wilholt draws attention to the parallels between this argument and that for decentralisation over centralised planning of the market.[9]

The arguments just surveyed motivate the funding of what we might call blue-skies research: basic science chosen for purely scientific reasons. Numerous anecdotes and empirical evidence show that many practical innovations stem from unrelated basic research conducted decades earlier.[10] Other empirical work has found that untargeted research and development (R&D) funding, in the form of public funding of research councils which set their own priorities, is strongly correlated with innovation, as measured by economic data on market sector productivity.[11] This same study found no correlation between market productivity and either targeted public R&D or public support for private R&D, such as tax credits.[12]

These studies provide empirical evidence to support the theoretical link between free, blue-skies research unrelated to national priorities and scientific progress. However, more directed research may also be necessary and justified in specific circumstances. As recently demonstrated by the COVID-19 pandemic, responding to events of overriding importance and urgency may require fast, targeted research efforts. A review of empirical evidence found that peer reviewers are reasonably successful in predicting the future success of proposed research, so long as the timescales involved are less than five years.[13] Thus, there are strong arguments in favor of funding short-term, directed research on urgent priorities that has been judged feasible and worthy of pursuit by peer reviewers.[14]

A third category of research funding, termed 'moonshot' or 'mission-driven' research, involves politically imposed priorities for abstract, long-term goals rather than immediate, short-term research. Examples range from successful initiatives like the actual lunar mission to less successful programs like Nixon's war on cancer.[15] This funding model combines elements of blue-skies and directed research, providing a time frame and flexibility for basic science to thrive, albeit within a space restricted by the envisaged goals. As is the case of directed research, mission-based research may be necessary to focus attention on particularly pressing national priorities, such as climate change. However, to the extent that they restrict scientific freedom and compete with blue-skies research, mission-based research programs can be expected to have restrictive effects on future innovation. Thus, mission-directed funding should only be used for the most important national priorities and should maximise the freedom of scientists to choose their own methods and approaches to achieving stated goals.

In line with the epistemological argument, the evidence presented here highlights the crucial role of scientific freedom in advancing knowledge and innovation. Blue-skies research, driven by pure scientific curiosity, is vital to scientific progress and should receive as much funding as possible. Nevertheless, there are specific scenarios where short-term directed and medium-term mission-based research can prove beneficial. These three categories of research funding—blue-skies, short-term directed and mission-based—are complementary, each serving human interests on different timescales. While blue-skies research should be prioritised, the other two types can be utilised in certain circumstances to address urgent needs. However, it is vital to minimise restrictions on scientific freedom in directed and mission-based research to ensure they do not detract from the essential role of blue-skies research in spurring innovation and enhancing human knowledge.[16]

Necessary for Democracy

A second justification for scientific freedom identified in the literature is known as the political argument.[17] The political argument posits that scientific freedom is necessary to sustain democratic legitimacy. This is because citizens often need to rely on scientific evidence to make up their minds concerning politically charged topics, such as welfare payments, climate change or genetic modification. Since an unbiased understanding of the facts is crucial to the legitimacy of a political opinion, science must remain free of political influence. Otherwise, the voters cannot trust that their political opinions are not being manipulated by those in power. This in turn would undermine the democratic legitimacy of the government trying to manipulate the science.

This justification was invoked in one of the first landmark cases on the topic, *Sweezy v. New Hampshire*, which concerned the prosecution of one Sweezy, a Marxist economist, for Communist subversion during the McCarthy era. Acting on sweeping powers conferred upon that office during this period, the Attorney General of New Hampshire issued subpoenas for notes and other material Sweezy had used in preparing and giving a lecture on Marxism, so that these materials could be used to establish Sweezy's views on the inevitability of communism and other related topics. Having refused to hand over these notes, Sweezy was held in contempt of court, a decision which he appealed all the way to the U.S. Supreme Court.

Although the Court found in favor of Sweezy on due process grounds, the judgement is notable for its emphasis on academic freedom:

The essentiality of freedom in the community of American universities is almost self-evident. No one should underestimate the vital role in a democracy that is played by those who guide and train our youth. To impose any strait jacket upon the intellectual leaders in our colleges and universities would imperil the future of our Nation. . . . Teachers and students must always remain free to inquire, to study and to evaluate, to gain new maturity and understanding; otherwise our civilization will stagnate and die.[18]

Implicit in this judgment are two distinct but related and oft-cited justifications for academic and scientific freedom: (1) its relation to democracy and (2) its necessity for education, and by extension the future interests of a population.

The link between academic and scientific freedom and democracy was noted by several surveyed studies. The claim can be advanced in either a weak or strong form. According to the strong version, true democracy is not possible, or not possible for long, without academic and scientific freedom, because the latter are necessary for the former. The weaker form simply states that freedoms are important, but not strictly necessary, factors in establishing and maintaining a strong democracy.

Required By International Human Rights Law

A third justification for scientific freedom raised in the literature was that it is required by international human rights law. Thus, Beiter and colleagues, accepting the distinction between academic and scientific freedom, identify Article 13 of the International Covenant on Economic, Social and Cultural Rights (ICESCR) as the 'natural home' of academic freedom, and Article 15 of the ICESCR as recognising a more general right to scientific freedom.[19]

Article 13(1) ICESCR recognises the right to education:

> The States Parties to the present Covenant recognize the right of everyone to education. They agree that education shall be directed to the full development of the human personality and the sense of its dignity, and shall strengthen the respect for human rights and fundamental freedoms. They further agree that education shall enable all persons to participate effectively in a free society, promote understanding, tolerance and friendship among all nations and all racial, ethnic or religious groups, and further the activities of the United Nations for the maintenance of peace.

The CESCR has interpreted this right as extending to the protection of academic freedom: '[i]n the light of its examination of numerous States parties' reports, [the Committee] has formed the view that the right to education can only be enjoyed if accompanied by the academic freedom of staff and students.'[20]

Scientific freedom is explicitly protected by Article 15(3) ICESCR, which states that '[t]he States Parties to the present Covenant undertake to respect the freedom indispensable for scientific research and creative activity.' However, to the extent that the epistemic justification for scientific freedom is correct, and scientific freedom is therefore required for scientific progress, it is also protected by Articles 15(1)(b) and 15(2) ICESCR:

> 1. The States Parties to the present Covenant recognise the right of everyone:
>
> (b) To enjoy the benefits of scientific progress and its applications
>
> 2. The steps to be taken by the States Parties to the present Covenant to achieve the full realisation of this right shall include those necessary for the conservation, the development and the diffusion of science and culture.

This conception of scientific freedom as required by international human rights law promises another level of protection to scientific freedom. Despite this promise, it has so far been neglected. This is likely due to the general neglect of the parent right to science. We explore the potential of scientific freedom under the right to science in greater detail in the following chapter.

IMPLEMENTING SCIENTIFIC FREEDOM

This cluster, which also involves seven thematic features, can be thought of as representing the difficulties raised in attempting to implement scientific freedom in practice. Many of these difficulties exist because scientific freedom can conflict with other values. In these cases, it is important to be clear about the justifications that ground scientific freedom, so that these (search for truth, democratic legitimacy, and obligations recognised by international human rights law) can be identified and balanced against other values (e.g., the need to protect society from harm or violations of their rights or dignity).

'Policy/Democracy' and 'Science V. Society'

Considering the literature, it appears that the political argument presents a strong case for scientific freedom in the context of political interference.[21] Thus, to retain democratic legitimacy, governments should refrain from politically motivated suppression of research results; in other words, they should *respect* the scientific freedom rights outlined in Article 15(3) and derivable from Articles 15(2) and 15(1)(b) of the ICESCR. The importance of this point is brought out by the large number of specific examples of governments interfering with scientific freedom for political reasons in our dataset of specific instances mentioned by included articles. The database contains more than 120 such examples and is appended to this book. Thus, included articles raised concerns about possible failures to respect scientific freedom by the Italian government in relation to restrictions on in-vitro fertilisation (IVF) treatments;[22] the American and German governments in relation to restrictions on recombinant DNA and stem cell research and technology;[23] the suppression of climate change and evolution science, primarily in the United States;[24] the denial of sexual education and contraceptives;[25] the taboo against studying racial or gender-based individual differences in intelligence;[26] travel bans for scientists based on their country of origin;[27] eugenics;[28] gene editing;[29] censorship;[30] Lysenkoism in the USSR;[31] excessive post-9/11 classification and scrutiny of research in the US;[32] human cloning;[33] synthetic biology;[34] partisan distribution of funding;[35] environmental impact statements[36] and institutional review boards (IRBs).[37]

Many of the examples of interference in scientific freedom by governments mentioned in the above list arguably have a purely political motivation. By contrast, a handful (e.g., environmental impact statements, research overview by IRBs, individual differences in IQ, eugenics, IVF treatments, evolution science, sexual education and contraceptives, human cloning, and synthetic biology) may have a political element, but are also motivated by other considerations. These include safety (environmental impact statements, censorship, research overview by IRBs, eugenics, human cloning, gene editing, post-9/11 classification and synthetic biology), religion (sexual education and contraception, evolution, censorship, human cloning, synthetic biology, IVF treatments and eugenics), morality (sexual education and contraception, human cloning, synthetic biology, gene editing, and eugenics and censorship), and dignity (IVF treatments, contraception, synthetic biology, gene editing and eugenics).

'Restrictions' and 'Responsibilities'

To identify which of these examples are worrisome from the perspective of scientific freedom, we need to understand these conflicts and 'tensions' and how they square with the underlying values of scientific freedom. This task is greatly aided by the fact that epistemological justification is bound up with human rights requirements. These in turn have a long history of scholarship and practice precisely addressing the issue of balancing one right versus another, and/or versus other, non-rights considerations. Making use of this pre-existing framework has numerous advantages. In addition to being required by international law and backed up by seventy-five years of experience and study, the framework for acceptable limitations on scientific freedom as a constituent of the right to science also boasts clear guidelines for resolving conflicts as well as identifying who should resolve them.

These limitation criteria, outlined in Article 4 ICESCR, are explored in greater detail in later chapters. Briefly, the limitation criteria state that scientific freedom may only be limited 1) according to the rule of law; 2) when strictly necessary to further the general wellbeing; and 3) in a way compatible with the nature of the rights contained in the Covenant. As we understand it in this context, this means when the limitation is necessary for the general well-being and compatible with the search for truth and the full development of the human personality. If this is the correct interpretation, it will have serious implications for the kinds of restrictions that governments can legally implement on scientific freedom without violating their human rights obligations, since few restrictions are likely to pass such a stringent test. Certainly, few of the examples of restrictions found in the surveyed literature come close to satisfying all three criteria.

However, this does not mean that there are no legitimate restrictions on scientific freedom. One such set of restrictions relates to scientific integrity and methodology. Although scientific freedom encompasses the freedom to try out new and various approaches, the underlying methodology must remain scientific. Otherwise, there can no longer be talk of a serious search for truth, and therefore no justification for scientific freedom. In making this point, Starck enumerates some of the many ways in which scientific integrity can be undermined:

Forgery and manipulation of data,

Selection and rejection of unwanted results,

Manipulation of a representation or an image,

Plagiarism and theft of ideas,

Assumption or unjustified adoption of scientific authorship or co-authorship,

Falsification of content,

Unauthorized publication by a third party, as long as the work, knowledge, hypothesis, or research program has not yet been published,

Sabotage of research activity through manipulation etc.,

Ignoring of primary data, to the extent that this breaches discipline specific recognized principles of scientific work.[38]

Indeed, these acts are 'inimical to science, seriously impair the scientific ethos, and cannot be included in an ethical conception of science which underlies the constitutional protection of freedom of scientific research.'[39]

In a 1996 judgement, a German Federal Court established that it could convene a committee to determine scientific misconduct and proceed accordingly, 'if and to the extent that serious objections can be raised against a scientist on the basis of concrete evidence, perhaps that he irresponsibly breached fundamental principles of science or abused research freedom, or that his work has to be denied the character of scientific.'[40] However, such objections cannot be made simply on the basis of errors, mistakes or unorthodox approaches, but only 'when it systematically lacks scientific character, not just occasionally or according to the views of certain schools of thought, such that in the light of its content and form one can no longer speak of a serious attempt to pursue truth.'[41]

This definition fits well with the Article 4 ICESCR limitation criteria: where one can no longer speak of a serious attempt to pursue truth, restrictions may well be necessary for the greater good of society. These are certainly compatible with the essence of the right to scientific freedom, which is to pursue truth.

Whereas restrictions imposed to prevent misconduct can easily be squared with the justifications underlying scientific freedom, the literature also mentioned several harder cases. For example, several studies touched upon the potential conflict between scientific freedom and IRB ethical oversight. Although meant to protect the health and dignity of those involved in medical experiments, included studies pointed out that IRB review requirements could be seen as amounting to licensing or censoring of science.[42] IRB review is discussed in more detail in Chapter Two of Part II.

An equally if not more important and contentious set of restrictions on scientific freedom stems from intellectual property (IP) protection. IP laws are theoretically supposed to incentivize innovation by making it possible for an inventor to obtain the exclusive rights to utilize an invention for some twenty years in the case of patents (and until seventy years after the author's death for copyright). Anyone wishing to make use of a patented or copyrighted work will then have to obtain a license to do so, usually involving a monetary payment and always subject to the discretion of the patent holder.

IP protections constitute a limitation to the extent that they take useful innovations out of the commons. This is a burden for many but is particularly acute for institutions and researchers in low- and middle-income countries.[43] Additionally, phenomena like patent thickets and patent trolls, as well as the granting of patents to whole groups of innovations and the evergreening of patents by minor modifications, all serve to remove useful innovations from the pool of resources available to other scientists.

'Dual Use'

The final major category of limitations mentioned in the surveyed literature were those meant to protect society and individuals against negative effects of science. As we saw in the opening vignette, the impact of some scientific advances depends crucially on how they are implemented and used. Nuclear physics, for example, may be used to build both power plants and warheads. This phenomenon, in which one application of a scientific concept or technology leads to beneficial, and another to deleterious, consequences, is known in the literature as the dual use problem. Tellingly, the phenomenon was first used to describe rocket science during the Cold War: the technology which could launch an astronaut into space is the same technology that launches intercontinental ballistic missiles. 'Dual use' since then refers broadly to any technology that can be used in the pursuit of two separate aims, but usually it is used in a narrow sense to describe technology with a civilian and a military use. The Internet, GPS systems, and many other modern civilian technologies are examples of originally military technology being used for civilian purposes.[44] The debate over treatment and enhancement in genetics and synthetic biology stems from the fact that advances in these fields might be used to improve biological traits in healthy, just as much as in sickly, individuals.

The surveyed studies contained numerous examples of the dual use phenomenon. These included nuclear technology;[45] gain-of-function studies with pathogens;[46] biosecurity/research involving pathogens more widely;[47] gene editing and synthetic biology.[48] Of this list, all except the latter two are instances of the narrow conception of dual use; the latter two are dual use technologies in the broader sense.

Restrictions on scientific freedom may be necessary to address dual use in safety-critical areas such as nuclear technology, given the enormous potential costs of misuse. Similar remarks can be made for gain-of-function and pathogenic studies promising the resurrection of the strains responsible for the 1918 flu epidemic and other deadly viral or bacterial strains.[49] This fear is articulated in the following quote from one of the included studies, which describes the publication of research showing how the H5N1 version can be aerosolised and thus made human-transmissible:

The benefits of publishing this work do not outweigh the dangers of showing others how to replicate it. . . . Someone might try to make it into a weapon . . . but a more likely threat is that more scientists will work with the modified virus, increasing the likelihood of it escaping the lab. Small mistakes in biosafety could have terrible global consequences.[50]

However, restrictions should only be implemented to the extent that they are strictly necessary and compatible with ICESCR rights. The necessity of reiterating this is borne out by a specific case studied by several of the included articles.[51] After the 9/11 attacks, the Bush administration introduced the Uniting and Strengthening America by Providing Appropriate Tools Required to Intercept and Obstruct Terrorism Act of 2001 (USA PATRIOT Act), Title VIII of which significantly modifies the rules surrounding pathogen research involving biological agents capable of being used as weapons. Specifically, the PATRIOT Act excludes wide categories of individuals from working with biological 'select entities' and imposes civil and criminal penalties for mishandling any such select entities. A 2010 evaluation of the effect of the PATRIOT and other relevant Acts found that, although they had not prevented research into select agents, they had reduced efficiency in the field, such that the same level of output required two to five times as much investment as previously.[52] Given that there are already very high costs involved in such research, and that this money cannot now be spent elsewhere in science, any such restrictions need good justification (the extent to which the PATRIOT and other Acts have prevented biological weapons misuse was not explicitly studied in the included articles).

'Government Intervention' Or Self-Regulation of Science?

Regulation of science does not have to stem from the government. The Asilomar genetic recombination meeting was mentioned by some studies as an example of scientific self-regulation.[53] The meeting took place at the Asilomar conference grounds in California in 1975 to discuss genetic recombination. Its immediate trigger was the anticipation of ecological danger, specifically biohazards, posed by advances in genetics. Those present at the conference managed to put into place a system of scientific self-governance, which was largely accepted by the broader scientific community. As part of that system, scientists voluntarily agreed not to conduct certain kinds of experiments considered too high risk, such as the cloning of highly pathogenic organisms or organisms carrying toxin genes. They also agreed to carry out less risky experiments under certain conditions meant to reduce the biological risks involved. This arrangement was accepted both by the scientific community, who appear to have voluntarily followed along with the consensus established at Asilomar, and by society and the government, who have allowed the scientific community to proceed under these conditions.

Indeed, a strong argument for scientific self-regulation is that it is scientists who best understand their area of expertise. This was illustrated in the included literature by the case of reactions to Dolly, the sheep cloned in 1996. Following the announcement of Dolly's birth, several US state legislatures rushed to enact legislation outlawing experiments involving cloning. Indeed, some twenty-six states enacted such legislation, using fourteen different definitions of cloning.[54] Five of these used a definition – 'growing or creation of a human being from a single cell or cells of a genetically identical human being through asexual reproduction' – which, if taken seriously, would criminalise the bringing to term of monozygotic twins (which arise from a single cell), and which more closely fits the type of cloning performed on plants than on humans.[55] More recently, an ordinance passed by Mencodino county in California, Measure H, which was aimed at banning the growing of genetically modified foods, defined DNA as 'a complex protein present in every cell of an organism . . . ' despite the facts that DNA is neither a protein (it is a polynucleotide rather than a polypeptide), nor is it present in all cells of an organism (blood, hair, and some skin cells do not have DNA).[56]

Adding to the difficulties presented by the fact that understanding the details of science is hard, the pace of science is faster than the pace of regulation; there is no guarantee that, by the time experts are brought in, compromises are made, and legislation is passed, the science will not have moved on to new methods or have novel implications. Thus, already in 1986, the US Office of Science and Technology noted that '[o]nce a relatively slow and ponderous process, technological change is now outpacing the legal structure that governs the system, and is creating pressures on Congress to adjust the law to accommodate these changes.'[57] It is well known that legislative progress may be slow. Legislators must consider very many issues, and they neither have the time nor the motivation or ability to fully comprehend the nuances of each issue or proposed solution to an issue. Their motivations include factors which may be politically necessary but are irrelevant or indeed harmful to an understanding of scientific issues on their own merits, and they are disproportionately affected by the way in which scientific issues are presented in the media and in public discourse.

In addition, at least in English-speaking jurisdictions, the common law system proceeds according to precedent. This necessarily introduces a level of conservatism and biases the judiciary away from grappling with novel questions on their own terms (as opposed to seeing them as extensions of previous technologies for which precedent is available, a strategy which does not do justice to the emergent properties of many such advances). This dynamic was illustrated by a case study in one of the included studies: when food additives 'found to induce cancer in man, or, after tests, found to induce cancer in animals,' were prohibited in the United States in 1958, it was thought that carcinogenic chemicals were quite rare.[58] Shortly thereafter, however, it surfaced that well over half of all chemicals (not only food additives) were carcinogenic if given in extreme quantities to laboratory animals, yet to possess only trivial risks in humans at the doses involved, and moreover that nearly every food additive as well as 'natural' food without additives contained some level of one of these chemicals. Nevertheless, the judiciary repeatedly struck down attempts by the FDA and other regulatory agencies to exempt substances with only trivial risks, on the basis that only Congress has the legislative authority to amend the relevant legislation; something which in the event did not happen until 1996.[59]

Another difficulty stems from the international, or universal, character of science. Science is universal in the sense that discoveries could be made or applied anywhere on the globe. Thus, if any given jurisdiction imposes new, restrictive regulations for a particular area of research, that area may still be investigated in other jurisdictions without restrictive legislation. In the words of the US National Research Council: 'Any serious attempt to reduce the risks associated with biotechnology must ultimately be international in scope, because the technologies that could be misused are available and being developed throughout the globe.'[60] This appears to have happened in the case of stem cell research, with some of the scientists involved moving away from countries like the United States and Germany, which have restrictive policies governing the research use of embryonic stem cells.

This last example also illustrates two further difficulties in the legislative regulation of scientific research. The first is the fact, already alluded to above, that whereas the point of departure for a scientist is the wish to generate evidence-based knowledge, there are numerous factors other than evidence-based knowledge which are salient in politics. Thus, the special-interest concerns of powerful constituencies may drive the political approach to a controversial area of science. Examples of this in the literature include the much-delayed approval of the morning-after pill by the FDA despite its safety having been convincingly demonstrated,[61] as well as any number of other reproductive and biological knowledge and technologies (IVF, sex education, synthetic biology, genetic engineering, contraception, evolution) to which pressures or biases arising from specific religious, moral or ideological viewpoints have prevented access despite scientific evidence of safety and efficacy. However, as evidenced by the opposition to GMO foods despite overwhelming evidence of their safety, these influences are not solely religious. Nevertheless, the point remains that parochial interests can hijack the political process whatever the science on the matter is.

In addition, despite its slow response to many scientific developments, the legislative process has been described as having an 'action bias':

> [a] tendency to take short-term action in response to a perceived immediate crisis while over-looking longer-term repercussions. The driving force seems to be the desire to take actions that will give the decision-maker the greatest and most immediate credit. The result of these forces is laws that may be superficially appealing, but which create many long-term problems and difficulties, a problem which has been referred to by one commentator as the pathology of symbolic legislation.[62]

Science regulation may be particularly prone to action bias since the complexity involved means that there is not much of a constituency to be pleased among the public. At the same time, though, this very complexity makes it easy to appeal to ignorance in fashioning criticisms of scientific research.[63]

What do these criticisms of legislative oversight of science mean for scientific freedom understood as a constitutive element of the right to science? Most importantly, they suggest that the scope for which legislative regulation imposes limitations on scientific freedom 'strictly necessary for the general wellbeing' is not large, and certainly does not include all of science. Indeed, they seem to indicate that alternatives, such as scientific self-regulation, may be better suited to promote the search for truth which justifies scientific freedom.

CONCLUDING REMARKS

What emerges from our systematic review of scholarly literature on scientific freedom is a heightened awareness of the crucial role played by scientific freedom—in and of itself, but also as a constituent part of the right to science.

In this chapter, we have sketched a coherent conception of scientific freedom from a human rights–based perspective. To do so we have relied on the results of a systematic review of academic literature. The results indicate that debates over scientific freedom have been going on for centuries. Sceptics have long questioned what it is about scientists or scientific inquiry that demands more respect and greater freedom than other human pursuits.

We identified three main justifications for scientific freedom in the reviewed literature. The first is an argument from truth, which states that to find truth, it is necessary to have the freedom to be wrong. The second is a political argument, which points out that science and scientific institutions play a key role in the political process and therefore must remain apolitical themselves. Finally, the third argument derives justification for scientific freedom from international human rights law and scholarship.

A human rights–based case for scientific freedom can be made from several human rights, including from health, freedom of speech or education. However, the most natural derivation in our estimation is from the human right to enjoy the benefits of the progress of science and its applications. Article 15(3) of the ICESCR directly talks about the obligation of States Parties 'to respect the freedom indispensable for scientific research and creative activity.' The Oxford Dictionary defines the word "indispensable" as 'absolutely necessary,' whereas its Cambridge rival defines the term as 'something . . . so good or important that you could not manage without it . . . '[64] These definitions both capture the idea that without scientific freedom, science would not progress, or not as quickly. This was a fundamental assumption in the literature, with nearly all included articles sharing this view. The idea is that scientific freedom is the price we must pay to enable the kind of reflective practices that are typically behind truly path-breaking discoveries.

Working with scientific freedom under the human rights umbrella has several distinct advantages. Firstly, UNESCO and the rest of the UN system work within the human rights paradigm when they issue soft law documents such as the ones we deal with in this book. Secondly, and perhaps more importantly, the human rights framework has had seventy-five years of maturation, with several of the key elements in a workable model of scientific freedom—what justifies it, what counts as legitimate restrictions, etc. – already worked out. The human rights–based approach allows us to draw on years of accumulated wisdom from similar areas and fields of human experience, using the same theoretical approach.

Under the conception of scientific freedom as a constitutive element of the right to science, the rules of the ICESCR and of international human rights law more generally apply when one is attempting to interpret the various rights and responsibilities of scientists. This allows debates to be filled with a certain concreteness. According to this conception, scientific progress is a human right in itself and instrumental to the fulfilment of several other, fundamental rights, and for these reasons scientific progress is valued highly. Scientific freedom is seen as indispensable to this progress and thus to the attainment of the benefits of scientific progress, the reason being that scientific progress is a creative process, and trial-and-error is fundamentally important for creative processes. A second reason is that it is relatively cheap to fund scientists compared to the value derived from their efforts.

Viewed in this way, scientific freedom quickly takes on moral dimensions. If your health depends on science, and science in turn depends on scientific freedom, then endangering scientific freedom leads to endangering people's health and well-being. But on the other hand, since we are working within a human rights framework, scientific freedom cannot be used to justify actions which themselves violate other human rights. Thus, a medical researcher cannot claim scientific freedom in experimenting on others without their consent. This naturally implies a balancing act between scientific freedom and the other interests and rights of those affected by scientific progress.

For all these reasons, the case for viewing scientific freedom as a constitutive element of the right to science is a strong one. Such a viewpoint makes it actionable and places it within a long tradition of scholarship and advocacy. Academically, the move can easily be defended, since scientific freedom is mentioned explicitly in Article 15(3) and is also necessary, as we shall see in the next couple of chapters, for the other parts of Article 15. Without scientific freedom, no scientific progress will be made from which everyone may benefit according to Article 15(1)(b).

Chapter 3

Taking Human Rights to the Next Level

The Right to Science, History and Content

The right to enjoy the benefits of the progress of science and its applications, in this book abbreviated to 'the right to science,' takes human rights to a new level. While several scholars around the world work on human rights today, very few focus on the right to science. In the twenty-first century, several of the defining challenges—from climate change and global pandemics such as COVID-19 to food security, the widening gap between rich and poor, and nuclear disarmament—have scientific dimensions. At the same time, the value of science has been under attack, with some raising alarm at the emergence of 'post-truth' societies and 'fake news'. As 'a philosophical idea, a legal promise, a political discourse, and—perhaps a social movement,'[1] the right to science is key to all this, notes Lea Bishop:

> Freedom of scientific enquiry. Integrity of research. Universal access to science education. Sharing knowledge and discoveries as a public good. Open access to academic research. International cooperation through scientific collaboration. Scientifically sound public policy. Broad participation in scientific enquiry. Attention to the needs of vulnerable communities. Widespread access to new technologies. Patent policy as a means to this end. Free and informed participation in experiments and new technologies. These are among the central themes of the internationally recognized human right to participate in the process of scientific discovery and to share in the benefits of technological progress—what has only very recently come to be shorthanded as 'the right to science.'[2]

In this chapter, we will explore scientific freedom as an integral aspect of the right to science. It enjoys a prominent position within Article 15 of the

International Covenant on Economic, Social and Cultural Rights (ICESCR) and is mentioned in its predecessor, Article 27(1) of the Universal Declaration of Human Rights (UDHR). The drafters of these documents considered scientific freedom essential to scientific progress and thus to the enjoyment of the right to science itself. For this reason, scientific freedom was elevated into a separate obligation in Article 15(3).

The origins of the right to science can be traced back to the first international human rights instruments, the American Declaration of the Rights and Duties of Man (ADRDM) and the UDHR. Adopted in 1948, these documents were created by the international community to express their moral and political commitments to all individuals, following World War II. They are notable for their inclusion of economic, social, and cultural rights in addition to civil and political rights. In both cases, the right to science is included in the broader category of cultural rights. Article XIII of the ADRDM phrases the right as follows:

ARTICLE 13

1. Every person has the right to take part in the cultural life of the community, to enjoy the arts, and to participate in the benefits that result from intellectual progress, especially scientific discoveries.
2. He likewise has the right to the protection of his moral and material interests as regards his inventions or any literary, scientific or artistic works of which he is the author.

The drafting of the ADRDM proceeded in parallel to that of the UDHR, although the former was ahead of the latter by a period of roughly six months. As a result, there were significant and bidirectional influences in their respective drafting processes. This can be appreciated by the similarity of the UDHR formulation to that of the ADRDM:

ARTICLE 27

1. Everyone has the right freely to participate in the cultural life of the community, to enjoy the arts and to share in scientific advancement and its benefits.
2. Everyone has the right to the protection of the moral and material interests resulting from any scientific, literary or artistic production of which he is the author.

Both the ADRDM and UDHR are Declarations: a category of political and legal instrument sometimes chosen by drafters and parties to indicate that their content is intended to be understood as aspirations or ideals, as opposed to legally binding commitments of the kind that would arise from formal treaties and conventions. A process to turn the rights contained in the UDHR into legally binding commitments was, however, embarked upon in parallel to the drafting of the UDHR. Due to its politically more sensitive nature and geopolitical developments such as the Cold War and de-colonisation, this process would not be completed until much later.

However, the adoption of the International Covenants on Economic, Social and Cultural Rights (ICESCR) and on Civil and Political Rights (ICCPR) by the United Nations in 1966 did finally transform the moral and political goals expressed in the UDHR into binding legal obligations. Article 15 ICESCR phrases the right to science as follows:

ARTICLE 15

1. The States Parties to the present Covenant recognise the right of everyone:
   a. To take part in cultural life;
   b. To enjoy the benefits of scientific progress and its applications;
   c. To benefit from the protection of the moral and material interests resulting from any scientific, literary or artistic production of which he is the author.
2. The steps to be taken by the States Parties to the present Covenant to achieve the full realisation of this right shall include those necessary for the conservation, the development and the diffusion of science and culture.
3. The States Parties to the present Covenant undertake to respect the freedom indispensable for scientific research and creative activity.
4. The States Parties to the present Covenant recognise the benefits to be derived from the encouragement and development of international contacts and co-operation in the scientific and cultural fields.

Due to its inclusion in the ICESCR, governments are expected to take measures to respect and ensure the right to science, just as they are expected to respect other fundamental rights such as freedom of speech and due process.

We start the present chapter by introducing the historical background to and drafting of the right to science. Next, we turn to the prominent role of scientific freedom within it. We explore scientific freedom in the drafting of the ICESCR and then look at scientific freedom in soft law instruments and scholarship. Its role is so integral to the right to science, we conclude, that it is possible to speak of scientific freedom as a constitutive element of the right to science. In other words, without scientific freedom, there is no right to science.

HISTORICAL BACKGROUND

The first known reference to the concept of a right to enjoy the benefits of scientific progress was made by Franklin Delano Roosevelt in his now-famous 1941 State of the Union Address, considered an intellectual precursor of the postwar international system.[3] Among the 'basic things expected of our people' as the 'foundations of a healthy democracy', Roosevelt included the 'enjoyment of the fruits of scientific progress in a wider and constantly rising standard of living.'[4]

The historical context of the right to science-related ideas in FDR's Four Freedoms speech is rooted in the social and political climate of the 1920s and 1930s. The 'Roaring Twenties' were marked by unprecedented postwar economic gains, partly facilitated by technological advancements, such as the railroad. This era, also known as the Gilded Age, brought prosperity, albeit unequally distributed.

The Great Depression brought the Gilded Age to an end, and the widespread poverty resulting from the economic downturn was believed to contribute to the rise of fascism and communism in Europe. Political leaders in the United States recognised the importance of science and technology in counteracting poverty and preventing a similar fate. Herbert Hoover stated that progress in the 1920s was 'due to the scientific research, the opening of new invention, new flashes of light from the intelligence of our people.'[5] FDR, elected in 1932, also recognised the significance of science and technology and campaigned on a platform of expanding the federal government's role in the economy to counteract the effects of the Great Depression.

By the later stages of the New Deal, the political climate became more favorable for science and technology. This shift resulted in the creation of various government programs and initiatives, such as the 1934 National Power Policy Committee, the Bankhead-Jones Act of 1935, the 1935 Social Security Act, and the 1937 National Cancer Institute. In 1938, a report characterised research as a 'national resource,' suggesting that it should be planned and utilised like other resources.[6] By 1940, science had become central to national concerns, particularly in preparing the military for potential conflict. 'Cumulatively, the qualitative changes in government science during the later New Deal presaged a new era even if war had not intervened.'[7]

FDR's 1941 Four Freedoms speech can be seen as an internationalisation of the New Deal.[8] The speech aimed to emphasise the importance of economic security, both domestically and internationally, in maintaining peace and stability. The reference to right to science ideas originated in its second draft, based on a five-page dictation by FDR and a longer, twenty-six-page draft by Adolph A. Berle, Jr., 'a believer in economic progress and scientific advancement.'[9] Berle believed that a functioning economy, serving all citizens, was necessary to counteract the rise of demagogues and authoritarians.[10]

The second draft contained the following phrase: 'the enjoyment of the fruits of technological progress in a constantly rising and widely diffused standard of living.'[11] This wording was later altered by FDR himself in the speech's fifth draft.[12]

The central idea behind the right to science-related sentence in the Four Freedoms speech is thus that science and technology are critical for improving living standards and material well-being. In turn, this economic security was seen as essential in combating the rise of fascism, communism, and revolution resulting from poverty and inequality.

The Four Freedoms speech by FDR marked a critical juncture in the recognition of the importance of scientific progress and economic security in maintaining peace and stability. The idea of including the benefits of scientific progress and its applications in a statement of universal rights, however, originated not from North America, but from *Latin* America. By the 1940s, several Latin American countries had recently adopted new democratic constitutions prominently featuring economic and social rights.[13] There was thus significant regional interest in the legal protection of human rights.[14] The failure of Latin American efforts to promote the importance of human rights in Allied conferences aimed at designing the postwar international system led to parallel efforts at establishing a regional system in which human rights were to be accorded a more prominent role.[15]

Building upon the insights from the Four Freedoms speech, the experiences of the New Deal era, and the Latin American context and legal culture, the drafters of the American Declaration of the Rights and Duties of Man were well positioned to acknowledge the significance of human rights in fostering international peace, security, and cooperation.[16] In their accompanying report to the ADRDM's first draft, they analyzed the development of international law and the role of human rights in fostering peace, security, and cooperation among nations.[17]

Prior to World War II, they pointed out, international law primarily focused on state sovereignty and allowed states to exercise complete control over their citizens. Humanitarian interventions were infrequent and only permitted in extreme cases of cruelty towards religious or racial minorities. However, the rise of totalitarian regimes and their suppression of basic human rights, such as freedom of speech, press, and assembly, made it evident that these violations could become a menace to global peace. The drafters noted that the denial of these rights facilitated the spread of false ideas about other nations, fostering animosity and aggression.[18]

The drafters emphasised two distinct perspectives on protecting human rights. First, ensuring fundamental rights like freedom of speech, press and access to information was necessary for effective cooperation between nations. When a population is denied these rights, they are unable to establish direct contacts, understand different perspectives, and ascertain the true intentions of other countries. Second, protecting human rights within each state is essential for developing free, self-reliant, and responsible individuals who can contribute to the international community.[19] The right to science as understood today contains elements addressing both perspectives.

Like Berle and Roosevelt before them, the ADRDM drafters envisioned an evolving conception of democracy and human rights which went beyond political freedoms, extending to social and economic aspects. In the accompanying report, the drafters emphasised the importance of organising the economic life of a state to ensure equality of opportunity and fair rewards for labor. Furthermore, they acknowledged the moral worth of each individual and the role of basic necessities like adequate nutrition, healthcare, housing, and sanitation in achieving one's full potential.[20]

In the text commenting on the draft right to science provision, the ADRDM drafters recognised an additional motivation behind a right to science provision—one essentially based on fairness. They noted that 'the opportunities for discovery and invention are the result of many generations of progressive effort, and that each generation is the heir of the civilization which preceded it and as such is entitled to share collectively in the benefit which its men of greater genius are able to draw from the conditions placed at their disposal.'[21]

During this formative period, the right to science centered on access to information, knowledge, and the tangible benefits derived from scientific innovations. The rationale for incorporating this right into international human rights law was twofold: to prevent political unrest stemming from dissatisfaction and to uphold fairness, as scientific achievements were viewed as part of the shared legacy inherited from previous generations. The legal culture of 1940s Latin America, which featured prominent economic, social, and cultural rights, further shaped the evolution of the right to science. This historical backdrop is crucial for comprehending the drafting of the right to science in both the UDHR and the ICESCR, which are explored in the subsequent section.

DRAFTING HISTORY

The right to science provision was discussed on four occasions during the drafting of the UDHR[22] and three times during the drafting of the ICESCR.[23] The drafting history of the right to science in UDHR revolves around several key themes, including access to and participation in science and its applications, the notion of scientific freedom, the balance between intellectual property protection and access and the role of science in society.[24]

In line with its historical precedents, and drawing explicitly on the draft ADRDM, the first discussions of the right to science by the drafters centered on access to the benefits of scientific progress in the form of knowledge, inventions, and discoveries.[25] The right to science provision had already at this early stage been placed together with culture and the arts, reflecting the drafters' understanding of science as part of the broader set of conditions necessary for humans to fully develop and express their personality. In line with this broader view of science and culture, discussions soon made clear that the right to science should be understood as covering not only passive benefits from scientific advancement, but also active participation in the process and conduct of science.[26]

Of particular relevance to current purposes was a proposed amendment by the Peruvian delegation to insert the word 'freely' into the draft right to science provision. The Peruvian delegation stated that, '[in] its opinion, not only must the right of every person to take part in the cultural, artistic and scientific life of the community be recognized, but also the right to do so in that complete freedom without which there could be no creation worthy of man.'[27] Thus, right from the start the importance of scientific freedom was linked to its instrumental value in fostering scientific progress.

None of the proposals and amendments discussed so far were seen as controversial. However, the inclusion of the text on moral and material rights of authors, which would eventually become Article 27(2) UDHR and Article 15(1)(c) ICESCR, as well as a proposed amendment by the USSR to include language stating that science should serve one or more purposes were much more controversial and generated significant debate.[28] Whereas the provision on moral and material rights would eventually be adopted, despite being defeated on several occasions, proposals to tie science to goals such as peace, international cooperation, and democracy were consistently rejected on the grounds that such aims could serve as excuses for unjustifiable interference by the state in science and its free development.[29] Several delegates argued that if science were to be tied to an aim at all, it should be the search for truth.[30]

International Covenant on Economic, Social and Cultural Rights

While the main inspiration for the right to science provision in the UDHR was the ADRDM, the drafting of the ICESCR was heavily influenced not only by the UDHR but also by the delegates and constitution of UNESCO. The text that served as the basis for discussion of the right to science provision in the ICESCR was provided by UNESCO, whose delegate made clear that it was intended to develop the right to science provision in the UDHR.[31] The discussions held during the drafting of the right to science provision in the ICESCR reflected this by largely touching on the same themes as those debated during the UDHR. In particular, efforts to include moral and material rights and one or more purposes for science were again by far the most controversial, and again resulted in the inclusion of the former but rejection of the latter.[32]

Notably, when introducing and defending the other provisions which make up the wider right to science—Article 15(2–4), and, to a lesser extent, 15(1)(c) – the drafters characterised these provisions as necessary for, or conducive to, scientific progress.[33] The inclusion of Article 15(2–4) is the primary deviation from, or expansion of, the UDHR's right to science provision. Whereas Article 15(1) ICESCR, with slight modifications in language, carries over the provisions of Article 27 UDHR, Articles 15(2–4) find no equivalent expression in the UDHR and are thus innovations with respect to that text.[34] The drafting history reveals that these additional measures derive from the UNESCO constitution and were seen as broad measures by which the rights introduced in Article 15(1) could be implemented by States in practice.[35]

Havet, the UNESCO representative, introduced the right to science provision in part by saying that the right to take part in cultural and scientific life presupposed individual initiative.[36] 'However,' he went on, 'as was stated in the Constitution of UNESCO, an effort was still required of the public authorities to promote such participation in cultural life and scientific progress, to encourage and co-ordinate activities to that end, to facilitate international exchanges, to relax the restrictions sometimes imposed by the State on cultural and scientific life, and, finally, to eradicate drastically all discrimination against individuals and groups.'[37] These elements are strikingly similar to those protected by Articles 15(2) and (4).

In its Article 1(1), UNESCO's constitution defines the purpose of the organisation, which is 'to contribute to peace and security by promoting collaboration among the nations through education, science and culture [. . .]'[38] Article 1(2) then provides specific measures for the organization to advance these purposes, including to 'maintain, increase and diffuse knowledge,' 'encouraging cooperation among the nations in all branches of intellectual activity,' and to 'recommend such international agreements as may be necessary to promote the free flow of ideas by word and image.'[39] Using the UNESCO constitution as a basis,[40] the drafters likewise set out a purpose for the right to science in what would become Article 15(1)(b) ICESCR—to recognise 'the right of everyone . . . to enjoy the benefits of scientific progress and its applications' – as well as specific methods of implementation in Articles 15(2–4).[41]

This reading is confirmed by a note on the general implementation article (Article 2 ICESCR) from the UN Secretary-General's office stating that the 'view prevailed that there should be a general article (article 2) containing what was felt to be the firmest commitment which could reasonably be undertaken in relation to all the rights treated in the covenant, but that its inclusion would not prevent the elaboration of what the obligation of the general article would signify in relation to any selected right, or even the imposition of stricter obligations in connexion with such a right. [. . .] article [15], paragraph 2, thus elaborate[s] upon the obligation of article 2 in relation to [. . .] rights relating to culture and science, while separate and additional obligations are included in [. . .] article [15], paragraph 3, on respect for the freedom indispensable for scientific research and creative activity.'[42]

SCIENTIFIC FREEDOM IN THE DRAFTING HISTORY OF THE ICESCR

The scientific freedom provision in Article 15(3) does not derive from UNESCO's constitution nor its delegates but rather from a proposal by the United States to replace the text that would become 15(1)(b) with the following:

The States Parties to the Covenant recognise the right of everyone: [. . .]

To enjoy freedom [*sic*] necessary for scientific research and creation.[43]

This proposal is notable for two changes. First, it introduced the wording 'recognise the right of everyone,' bringing the right to science provision in line with other proposed rights and making clear that it, like these other rights, was intended to express a legally binding obligation. The second addition involved a proposed text addressing scientific freedom, which would eventually become Article 15(3) ICESCR. However, this latter change had replaced the previous provision, now known as 15(1)(b) ICESCR, instead of supplementing it. The reactions by other delegations to this proposal are interesting because they highlight the value and importance placed by them on scientific freedom.

Introducing the US proposal, Eleanor Roosevelt explained that '[e]mphasis had been laid upon the freedom necessary for scientific research and creation because the original text called merely for the right to enjoy the benefits of scientific progress, or, in other words, simply the right to enjoy the results of scientific research, whereas what was really required was to ensure conditions in which such research could be freely conducted.'[44] In response, Valenzuela of Chile agreed on both the importance of making the right to science binding and introducing a provision protecting scientific freedom, but argued that instead of one replacing the other, both ought to be recognised. 'He therefore urged the United States delegation to retain [Article 15(1)b] and, incidentally, to delete the word 'necessary' in [the suggested text for Article 15(3)], which in the present context might have a restrictive meaning.'[45]

The Lebanese delegate responded to this exchange by suggesting a sub-amendment to the US amendment, 'contain[ing] a new paragraph 3 to be added to article 30 to replace sub-paragraph (b). That new paragraph would provide that States Parties to the covenant undertake to respect the freedom indispensable for research and scientific invention.'[46]

The view of scientific freedom as a prerequisite for the progress or advancement of science was most explicitly expressed in the reactions of the delegates to repeated proposals to subject the development of science to goals and purposes. Bey of Egypt asserted, for example, that 'science should be free from any mundane considerations, no matter how praiseworthy, and scientists should receive no guidance from outside but should obey only their own conscience and the exigencies of their work. The search for truth must remain unshackled.'[47] Supporting this, Hoare of the UK 'entirely agreed with the Egyptian representative concerning the USSR amendment. Science in the past had always grown from within; it was and must remain autonomous, and no external criterion, no matter how praiseworthy, should be applied to it or to its development.'[48]

Early Cold War concerns were obvious in the responses by Hoare as well as by several delegates from around the world who expanded on these arguments. Though various delegations found the idea behind the Soviet proposals important, they were defeated because they were seen as an attempt to use science to further a particular political ideology. Thus, Australia's Whitlam maintained that '[s]cience could be regarded only as an autonomous growth, and as such should not be subjected to any interests, however admirable they might be in themselves.'[49] '[S]cience and culture could develop only in an atmosphere of complete freedom,' added Chaudhury of Pakistan, and continued: 'The State should not therefore impose restrictions on scientific research or control creative activity, but on the contrary eliminate all obstacles.'[50] As UNESCO's Maheu saw it, 'by their very nature, either [science, education, and culture] were free or they did not exist'[51] – a sentiment that was echoed by India's D'Souza who stated that 'certainly scientific and cultural progress was conceivable only in a climate of freedom.'[52]

Repeated, failed attempts to attach a purpose to scientific progress was the only exception to the otherwise general endorsement of the importance of scientific freedom. Debates on 15(3) centered instead on the inclusion of a single word, 'indispensable.' The controversies surrounding the inclusion of the word 'necessary' and later 'indispensable' concerned whether these should be understood restrictively, as limiting the scope of scientific freedom protected to that which is necessary or indispensable for scientific progress; or expansively, as stating that scientific freedom *tout court* is necessary or indispensable for scientific progress and must therefore be respected.

Several delegates, as well as the Secretary-General of the UN, noticed the potential for a restrictive reading according to which only some, but not all, freedom is necessary or indispensable for scientific progress and opposed the inclusion of these words on these grounds.[53] Brillantes of the Philippines said, for example, that "[i]n paragraph 3 of the article, the use of the word 'indispensable' gave the impression that the State undertook only to respect a strict minimum of freedom necessary for scientific research and creative activity. He doubted that that would give creative activity much encouragement. Moreover, it was the State that would determine the degree of freedom considered indispensable. It was therefore possible that paragraph 3 might have the effect of limiting or nullifying the scope of paragraphs 1 and 2..'[54] Agreeing with Brillantes, Hastad of Sweden added that "the word 'indispensable' in paragraph 3 of the article might lead to confusion. In view of the fact that it added nothing to the text that was not already implicit in it, it might be advisable to delete the word.'[55]

To resolve the issue, the Philippine delegation, backed up by Guatemala, requested a separate vote on the inclusion of the word 'indispensable.'[56] Should it come to a vote, said UK delegate Hoare, who strongly disagreed with his colleagues from the Philippines and Sweden, 'he would vote to retain it.' He explained his wish to keep the word in the final text by referring to the fact that 'the Commission on Human Rights had inserted the word deliberately and after careful consideration.' He then proceeded to remind everyone that

> where the relations between the State and scientists and artists were concerned, [. . .] no State was in a position to allow absolute freedom; the restrictions imposed by the requirements of public order and national security were inevitable, particularly with regard to scientific research. If the word 'indispensable' were deleted, the State's capacity to impose limitations on the activities of scientists and artists as citizens might be placed in doubt. The Commission had recognized that the core of the matter was the kind of freedom that was indispensable for scientific research and creative activity and that the provision should be directed to the protection of that freedom.[57]

Hoare's argument carried the day. In the end, it was decided by 41 votes to 9, with 23 abstentions, to retain the word.[58] Summing up the various arguments made, the drafting history implies that the scientific freedom protected by Article 15(3) is that which is a) necessary for scientific progress, in the sense of contributing in some essential way to that progress; and b) does not threaten public order, national security, the general well-being, and similar interests.

What the drafting history furthermore suggests is that for many drafters, the scientific freedom protected by Article 15(3) implies negative, not positive, freedom. Responding to the proposed US amendment, the UK delegate 'wondered whether States would thereby assume the obligation to provide the freedom necessary for scientific research and creation, say by endowing scientists and artists with a private income. While he agreed with the idea expressed in its most general sense, he felt that the provision should be made clearer lest it should be interpreted in that manner.'[59] During the same debate, the Lebanese delegate stated that 'the right to participate freely in cultural life and to engage in scientific research did not imply the extension of facilities and time to everyone for those purposes, but rather the fact that no individual would be prevented from pursuing those activities.'[60]

Scientific Freedom as a Constitutive Element of the Right to Science

If Article 15(3) itself, when read in isolation, suggested negative freedom to many drafters, a somewhat different picture emerged once scientific freedom was seen as closely related to the other parts of Article 15, and to the right to science in general. Perceiving scientific freedom to be of crucial importance for the right to science, these drafters did not only consider it worthy of protection in its own right. The drafting history shows that they also deemed scientific freedom worth defending because they saw it as instrumentally necessary for advancing scientific progress. There can and will be no scientific progress without scientific freedom; or, to put it in a different way, scientific freedom is a constituent, or integral, element of the right to science: one of the pieces of which this right is made up, and without which it would no longer be the right to science.

This insight is interesting because it affects the degree and extent to which international human rights law can be interpreted to protect scientific freedom. Looking at the wording of Article 15 ICESCR and of the drafting history pertaining to Article 15(3) in isolation, one might plausibly argue that their protection of scientific freedom extends only to negative scientific freedom: that international human rights law may be said to protect against government interference in science but does not require positive investments in and facilitation of science. Moreover, even this negative right could plausibly be interpreted as limited by relatively broad state interests such as public order, national security, and the like, as indicated by UK delegate Hoare above.

Viewing scientific freedom as a constitutive element of the right to science, however, allows for a broader interpretation of the level and scope of protection of scientific freedom under international human rights law. To the extent that (positive) scientific freedom is indispensable for scientific progress to occur, (positive) scientific freedom can be said to be required not through Article 15(3), but rather through Article 15(1)(b) itself. Furthermore, if scientific freedom really must be protected for scientific progress to occur, then it would seem also to be necessary for carrying out the State obligation in Article 15(2), viz., to conserve, develop and diffuse science. Likewise, both positive and negative scientific freedom may be required to fulfill the aspirations envisaged by Article 15(4) of a world in which scientists are free to travel and exchange their views, equipment and data.

SCIENTIFIC FREEDOM IN SOFT LAW INSTRUMENTS

Although the drafting history of the right to science contains perhaps the clearest expression of the rationale for protection of negative and positive scientific freedom under international human rights law, the interpretation advanced here of scientific freedom as a constitutive element of the right to science is backed up by the importance attached to scientific freedom in subsequent soft law instruments and scholarship pertaining to the right to science. It likewise receives support from the other soft law instruments within the broader normative environment of the right to science, as well as from state practice and national constitutions. The most important soft law instruments are the 2009 Venice Statement on the Right to Enjoy the Benefits of Scientific Progress and its Applications, the 2012 Report by the Special Rapporteur in the Field of Cultural Rights, Farida Shaheed, and the 2020 General Comment No. 25 on Science and Economic, Social and Cultural Rights. In the below sections, we survey the treatment of scientific freedom in each of these instruments. The General Comment is of particular importance and is examined in depth in Chapter Two of Part Two, as well as in Chapter One of Part Three, in which one of its primary authors, Mikel Mancisidor, provides an account of its development and content.

The Venice Statement

The Venice Statement on the Right to Enjoy the Benefits of Scientific Progress and Its Applications resulted from three expert meetings on the subject, sponsored by UNESCO.[61]

This statement acknowledges the promotion of science, participation in science, scientific freedom, access to benefits and protection from harms as components of the normative content of the right to science:

The normative content should be directed towards the following:

a. Creation of an enabling and participatory environment for the conservation, development and diffusion of science and technology, which implies *inter alia* academic and scientific freedom [. . .]
b. Enjoyment of the applications of the benefits of scientific progress, which implies *inter alia* non-discriminatory access to the benefits of scientific progress and its applications [. . .]
c. Protection from abuse and adverse effects of science and its applications.[62]

In line with the understanding of scientific freedom as contributing to scientific progress advanced above, the Venice Statement goes beyond identifying scientific freedom as part of the normative content of the right to science by linking freedom of inquiry to the development of science more broadly: 'freedom of inquiry is a vital element in the development of science in its broadest sense. Science is not only about advancing knowledge of a specific subject matter, nor merely about procuring a set of data and testing hypotheses that may be useful for some practical purpose. It is also, at the same time, about enhancing the conditions for further scientific and cultural activity.'[63]

In its treatment of State obligations under the right to science, the Statement again links scientific freedom to scientific progress:

The duty to respect should include:

a. to respect the freedoms indispensable for scientific research and creative activity, such as freedom of thought, to hold opinions without interference, and to seek, receive, and impart information and ideas of all kinds.[64]

Consonant with our understanding of scientific freedom as a constitutive element of the right to science, the Venice Statement asserts that the right in Article 15(1)b ICESCR is 'inextricably linked' to scientific freedom as recognised by Article 15(3) ICESCR.[65] It furthermore links scientific freedom to international cooperation, 'including the freed [*sic*] exchange of information, research ideas and results.'[66]

The Special Rapporteur's Report

The 2012 report by the Special Rapporteur in the field of Cultural Rights on the right to science adopts much of the Venice Statement's language and incorporates essentially the same elements in its outline of the right's normative content, with the addition of two forms of participation (engagement in science and involvement in decision-making related to science):

> The normative content of the right to benefit from scientific progress and its applications includes (a) access to the benefits of science by everyone, without discrimination; (b) opportunities for all to contribute to the scientific enterprise and freedom indispensable for scientific research; (c) participation of individuals and communities in decision-making; and (d) an enabling environment fostering the conservation, development and diffusion of science and technology.[67]

The report explicitly endorses the Venice Statement on the link between scientific freedom and scientific progress, emphasising that 'freedom of inquiry is vital for advancing knowledge on a specific subject, procuring data and testing hypotheses for some practical purpose, as well as for promoting further scientific and cultural activity.'[68] It furthermore argues that scientific freedom has a participatory aspect, such that '[f]reedom of scientific research includes the right of everyone to participate in the scientific enterprise, without discrimination on the basis of race, colour, sex, language, religion, political or other opinion, national or social origin, property, birth or other status.'[69]

Of particular interest in the current context, the Rapporteur links scientific freedom not only to the right to science generally, but specifically to Article 15(2) ICESCR. As argued above, scientific freedom is crucial for each of the obligations in Articles 15(2–4), which were intended as specific measures of implementation to facilitate the fulfilment of the right to science as expressed in Article 15(1)(b). Article 15(2) enumerates three specific implementation measures: the conservation, development and diffusion of science. Addressing these, the Special Rapporteur argues that

> [d]evelopment demands an explicit commitment to the development of science and technology for human benefit by, for example, developing national plans of action. Usually, this implies the adoption of programmes to support and strengthen publicly funded research, to develop partnerships with private enterprises and other actors, such as farmers in the context of food security, and to promote freedom of scientific research.[70]

Finally, the report stresses that scientific freedom, like the right to science and human rights generally, is not absolute but may be subject to limitations under the clearly defined criteria laid out in Article 4 ICESCR.[71]

The General Comment

The 2020 General Comment No. 25 on the right to science likewise identifies access and participation rights, scientific freedom, and the promotion of science as aspects of the right's normative content.[72] It is also explicit in tying scientific freedom to scientific progress: 'In order to flourish and develop, science requires the robust protection of freedom of research.'[73] The General Comment goes on to analyze the components of scientific freedom in the following way:

> This freedom includes, at the least, the following dimensions: protection of researchers from undue influence on their independent judgment; the possibility for researchers to set up autonomous research institutions and to define the aims and objectives of the research and the methods to be adopted; the freedom of researchers to freely and openly question the ethical value of certain projects and the right to withdraw from those projects if their conscience so dictates; the freedom of researchers to cooperate with other researchers, both nationally and internationally; and the sharing of scientific data and analysis with policymakers, and with the public wherever possible.[74]

Like the Special Rapporteur's report, the General Comment explicitly connects scientific freedom to the obligations to conserve, develop, and diffuse science in Article 15(2) ICESCR: 'States parties should not only abstain from interfering in the freedom of individuals and institutions to develop science and diffuse its results. States must take positive steps for the advancement of science (development) and for the protection and dissemination of scientific knowledge and its applications (conservation and diffusion).'[75] The language used makes clear that the Committee sees these obligations as extending beyond negative to positive duties, a reading confirmed by Mancisidor in his chapter in the present volume.[76] This is confirmed in paragraph 15 of the General Comment:

> The right to participate in and to enjoy the benefits of scientific progress and its applications contains both freedoms and entitlements. Freedoms include the right to participate in scientific progress and enjoy the freedom indispensable for scientific research. Entitlements include the right to enjoy, without discrimination, the benefits of scientific progress. These freedoms and entitlements imply not only negative, but also positive obligations for States. [. . .][77]

The General Comment makes further mention of positive obligations in relation to scientific freedom in its section on specific obligations under the right to science. The Committee here and elsewhere adopts the widely used tripartite typology, which classifies State obligations under the three headings of 'respecting,' 'protecting,' and 'fulfilling' a right. To respect a right means that a State party must not itself infringe it; protecting it goes beyond this in obliging State parties to prevent third parties from violating the right; and, most demandingly, fulfilling a right means bringing about the conditions necessary for its full enjoyment. The latter two types of obligations can be analyzed as containing positive obligations.

As an example of obligations to protect the right to science, the General Comment lists 'ensuring that private investment in scientific institutions is not used to unduly influence the orientation of research or to restrict the scientific freedom of researchers.'[78] After linking the obligation to fulfil the right to science to Article 15(2) ICESCR, the Committee makes clear that,

> States parties not only have a duty to allow persons to participate in scientific progress; they also have a positive duty to actively promote the advancement of science through, inter alia, education and investment in science and technology. This includes approving policies and regulations that foster scientific research, allocating appropriate resources in budgets and generally creating an enabling and participatory environment for the conservation, the development and the diffusion of science and technology. This implies, inter alia, protection and promotion of academic and scientific freedom, including freedom of expression and freedom to seek, receive and impart scientific information, freedom of association and freedom of movement; guarantees of equal access and participation of all public and private actors; and capacity-building and education.[79]

Like the Special Rapporteur, the Committee emphasises that scientific freedom is not absolute. However, it makes clear that 'any limitation on the content of scientific research implies a strict burden of justification by States, in order to avoid infringing freedom of research.'[80] The importance of any such limitations adhering to the Article 4 ICESCR criteria is underlined by the inclusion in the General Comment, among the minimum core obligations under the right to science, the removal of 'limitations to the freedom of scientific research that are incompatible with Article 4 of the Covenant.'[81] As examples of potential limitations to scientific freedom, the Committee highlight the precautionary principle[82] as well as citizen and public participation in scientific decision-making.[83]

Of direct relevance to our characterisation of scientific freedom as a constitutive element of the right to science, the Committee makes it clear that 'the protection of freedom of scientific research is also an element of the right to participate in and to enjoy the benefits of scientific progress and its applications.'[84]

SCIENTIFIC FREEDOM IN THE RIGHT
TO SCIENCE SCHOLARSHIP

During its initial fifty years, the right to science was scarcely discussed in academic circles. A few pioneering studies on the subject were conducted by researchers with a broader interest in human rights, science, and technology. In 1983, ICJ Justice C. G. Weeramantry published *The Slumbering Sentinels*, examining potential human rights consequences arising from advancements in information and communications technology and biotechnology.[85] Richard Pierre Claude's 2002 book, *Science in the Service of Human Rights*, was the first to concentrate solely on the right to science.[86]

The turn of the millennium marked a growing interest in the right to science. A few publications exploring the right to science and intellectual property (IP) emerged, potentially influenced by the then-recent TRIPS Agreement and other IP protection advancements.[87] Between 2009 and 2010, two comprehensive legal examinations of the right were published, leading to a noticeable increase in academic curiosity.[88] Although primarily focused on IP, these articles attracted a wide readership and numerous citations from scholars keen on applying the right to science to other fields. Their release coincided with the 2009 Venice Statement.

The 2012 report by the Special Rapporteur in the field of Cultural Rights on the right to science further piqued scholarly interest. A thorough review spanning from 1998 to 2020 identified a total of seventy articles and books discussing the right to science. This review highlighted the most frequent themes, such as scientific freedom, access, participation and inclusion, conflicts with IP protection, dual use, and balancing or relating to other rights.[89] The publication of General Comment No. 25 on the right to science in 2020 marked a turning point in the scholarship on this subject, which was also the year the first edited volume on the topic was published.[90]

In addition to soft law instruments, scientific freedom has been addressed by scholars working on the right to science. The systematic review of the right to science scholarship mentioned above found academic or scientific freedom mentioned in forty-four out of the seventy studies included as of 2020,[91] though none focuses primarily on scientific freedom.

Since 2020, several articles of direct relevance have been published. Yotova and Knoppers reviewed State practice relevant to the right to science as a means of ascertaining how States have themselves interpreted their right to science obligations.[92] Under Articles 16 and 17 ICESCR, State parties are required to submit reports to the Committee on Economic, Social and Cultural Rights in which they detail the steps they have taken to fulfill the rights therein.[93] Yotova and Knoppers looked at these reports, as well as responses to a questionnaire circulated by the Special Rapporteur in the field of Cultural Rights in preparation for her 2012 report. Since part of the Committee's guidelines explicitly ask State parties to report on 'measures taken to guarantee the freedom of exchange of scientific [and] technical information,' many of these responses are relevant to scientific freedom in the context of the right to science.[94] Of the 123 State parties reporting on their implementation of the right to science, 41 mention legislation guaranteeing the freedom of scientific research, making scientific freedom the most frequently mentioned topic in State reporting on the right to science. Of these forty-one, half also mention taking positive steps to promote science; and overall, thirty-nine States report doing so. Based on their survey, Yotova and Knoppers conclude 'that the right's core content should be interpreted as including freedom of scientific research, and an obligation on the state to enable access to scientific information and to support the development of science.'[95]

Boggio and Romano have carried out a review of right to science elements in national constitutions, among which they include academic and scientific freedom.[96] Breaking down constitutions by geographical region, and with the caveat that because 'a comparative analysis of all freedoms is beyond the scope of this entry, we have identified a few notable ones,' they identify five countries in the Americas, 15 in Africa, 'several' in the Middle East, of which three are given as examples, 23 in Europe, and five in Asian and Pacific countries that protect scientific and academic freedom.[97]

In a conceptual mapping of the contents of the right to science, Boggio links scientific freedom to scientific progress, stating: 'the public's entitlement to scientific progress joins with the freedom of scientists to be able to "do science" [. . .] unambiguously in the text of Article 15(3), which provides that State parties must 'respect the freedom indispensable for scientific research.'[98] Boggio analyzes the resulting obligations in line with the tripartite typology into 'three steps: respecting the scientific enterprise's autonomy; protecting its enterprise from undue external influences; and creating a supportive environment.'[99]

Finally, Beiter provides an extensive and in-depth analysis of scientific and academic freedom in relation to the right to science, which is cited and quoted at various places in this book.[100]

CONCLUDING REMARKS: FACILITATING
SCIENTIFIC PROGRESS FOR THE BENEFIT OF ALL

More broadly, the history and drafting of the right to science indicate that the drafters had a coherent and plausible view of the role of science and its applications in society. According to this view, the progress of science and its applications is important, due to the inherent value of the advancement of knowledge. Perhaps more importantly, this advancement is attended by practical benefits in the form of technology leading to better living standards, as well as knowledge that is useful for the understanding and attainment of goals, whether of a political, societal, ethical or personal nature.

The progress of science and its applications are consequently worthy of protection along with participation in and enjoyment of society, culture, and the economy more widely, because all these elements of the human endeavor serve the common purpose of contributing to the greater ability to fully develop and express the human personality. This latter purpose has been described as 'a way of summarizing all the social, economic, and cultural rights' in the UDHR and, by extension, all the rights in the ICESCR.[101]

Thus, one purpose of the right to science is to facilitate the progress of science for the benefit of everyone. For this to happen, however, several conditions must be in place: prime among these is that the processes of science must actually take place. These processes in turn require the presence of goods such as education and resources, but also require the absence of inhibiting factors such as political oversight and other forms of external control. In short, for humans to benefit from the knowledge and practical means necessary to fully develop themselves along their chosen paths, the science that creates such knowledge and means must progress, and for that to happen, it must be free. Hence the integral role of scientific freedom for the realisation of the right to science.

In Part Two, we develop this insight into a model of Scientific Autonomy and Freedom as an Integral part of the Right to Enjoy the benefits of Science: SAFIRES. This model in turn serves as our basis for evaluating international soft law instruments. But first, in the fourth and last chapter of this part of the book, we turn to the relationship of scientific freedom to the dissemination of science and the importance of international cooperation in the field of science, as outlined in Articles 15(2) and 15(4) ICESCR, respectively.

Chapter 4

Scientific Freedom

How Does It Relate to Scientific Dissemination and International Co-operation?

Scientific freedom is a constitutive element of the right to science, as we have seen. Not only is it specifically protected in Article 15(3) of the International Covenant on Economic, Social and Cultural Rights (ICESCR); without scientific freedom, no scientific progress would be made from which everyone may benefit according to Article 15(1)(b). Furthermore, in addition to respecting scientific freedom, States Parties to the ICESCR are obligated, according to Article 15(2) to take steps towards conserving, developing, and diffusing science. They must also recognise the importance of international contacts and co-operation in the scientific field, as outlined in Article 15(4).

Among the positive steps to be taken by States Parties to fulfil the right to science the 2017 UNESCO Recommendation on Science and Scientific Researchers mentions the promotion of research and technology, funded by public, private, and nonprofit sources, as well as the strengthening of scientific culture, public trust and support for the sciences throughout society.[1] At the core of promoting science as a public good is a properly trained and adequately supported scientific workforce. The open communication of the results of this workforce provides the best promise of scientific accuracy and impartiality, "as suggested by the phrase 'academic freedom.'".[2]

Access to research results is also a precondition for public participation in democratic processes to determine the appropriate uses, and possible limits of, scientific research and its application. When the steps taken towards conserving, developing, and diffusing science are combined, Article 15(2) arguably 'lays out a concept of access that connects state support for science (development) with the equitable distribution of the benefits and applications of scientific progress (diffusion) and efforts to ensure that these benefits are lasting (conservation).'[3] In previous chapters, we have discussed how scientific freedom relates to the positive steps to be taken by States Parties to fulfil the right to science in relation to the development and conservation of science. In the present context, we will instead zoom in on the issue of scientific freedom as it connects to the dissemination part of Article 15(2) as well as to the issue of international cooperation in the scientific field, mentioned in Article 15(4).

This chapter is divided into three parts. In the first two parts, we look at the importance of scientific communication and dissemination for the exercise of scientific freedom and for the success of the right to science in general. While it is entirely clear that the public cannot exercise their right to science unless scientific knowledge is disseminated in such a way that they understand it, it is not quite obvious to whom the task of communicating this knowledge falls. Should scientists themselves be prepared—perhaps even be educated—to let the public know the results of their research, or are professional science journalists better equipped to do so? What venues should be used for such communication in order that the interest in science may be furthered and trust be regained at a point in time where many have a (sinking) feeling of living in a 'post-truth' world in which scientific facts commonly agreed upon by scientists are routinely reduced to 'fake news' or 'alternative facts'? These are some of the questions with which we engage in the first part of this chapter.

In the second part, we look at the role citizen scientists could play in all this. We also debate how scientists can learn how to engage with the public and to honor Article 15(2) and its demand for scientific dissemination in a respectful, non-paternalistic way. Furthermore, we discuss the chance that communicating with the public gives to scientists of creating support for their research. If, as some would argue, the very fact that access to and participation in science and in policymaking concerning matters relating to science is part and parcel of the right to science in a democratic society, then it is of vital importance for scientists to show the public why their particular line of research deserves to be funded. Communication between the world of science and the public is not a one-way street.

As much as access to scientific knowledge generated and the applications in which such knowledge results are indispensable elements of the public's right to science, it is also crucial for scientists themselves to have access to the scientific knowledge produced by and discussions engaged in by their colleagues around the world. This is where Article 15(4) enters the picture. The third part of this chapter discusses the issue of openness and collegial sharing in the world of science. What limitations apply to openness—in order, for example, to anticipate dual-use science and technology or for reasons of national security or of privacy?

This and related issues have been discussed in the context of science diplomacy. During the Cold War, for example, contacts or interactions and exchanges within the scientific community were substantially limited, and this had a serious effect on scientific innovation both locally and internationally.[4] More recently, after the Russian invasion of Ukraine in February 2022, scientists in the West have had a difficult time working with Russian colleagues. What does that do to the openness and international cooperation that is so crucial to scientific progress—and to the scientific freedom which is at the heart of the right to science?

It is only, we argue, when all four parts of Article 15 ICESCR play together that scientific progress and its applications become the global common good envisaged by the drafters of both the 1948 Universal Declaration of Human Rights (UDHR) and the ICESCR from 1966. As we have seen throughout this book, it is primarily, though not exclusively, the perspectives of the scientific community that are in focus. Without inspiration and input from fellow scientists, citizen scientists, and others around the world, individual scientists and their research groups will have none of those original ideas that lead to progress, short- or long-term. Their scientific freedom will not result in the pursuit of new ideas. There will be nothing to disseminate (and to conserve and develop) – and therefore nothing for the public to benefit from and share in the end.

TRUST AND THE DISSEMINATION OF SCIENCE

On 1 August 1922, the International Committee on Intellectual Cooperation (ICIC) of the League of Nations met for the first time. On 21 September the year before, a resolution had been adopted by the second assembly of the League to create a consultative body whose purpose would be 'to examine international questions regarding intellectual co-operation, . . . to consist of not more than 12 members and to contain both men and women.'[5] The League of Nations had itself been founded in 1920 by the Paris Peace Conference that ended the First World War, and its main mission was to 'promote international co-operation and to achieve international peace and security.'[6]

Some of the delegates to the first assembly thought that the League could never succeed in this mission if intellectual work was not paid as much attention as that enjoyed by manual labor. A letter was therefore sent out to member nations asking for suggestions of eminent intellectuals who could sit on the ICIC. Among those who made it to the final list of twelve members were scientists Albert Einstein and Marie Curie; French philosopher Henri Bergson became the first Chairman. Initially, things did not quite work out. Einstein resigned after just one year on the ICIC, for example, because he did not think that the League operated in an efficient manner. But then he seemingly changed his mind, writing in a confidential letter to Vice Chairman of the ICIC Gilbert Murray in May 1924 that, 'whatever the failures of the League of Nations in the past, it must be regarded as the one institution which holds out the best prospect of beneficient action in these sad times.'[7]

As a consultative body, the ICIC was free to establish and pursue its own agenda, but the League did ask its members to look into three particular issues: 'how to improve the international organization of scientific research; how to establish relations between different universities; [and] how to organize scientific information systems and the exchange of publications.'[8] With the founding of the International Institute of Intellectual Cooperation (IIIC) in Paris in 1926, the Committee grew to become a real center of activity and over the next nearly twenty years, the idea of intellectual cooperation inspired the work of other bodies and institutions set up at the end of the First World War to create a new world order based on multilateral cooperation. It ceased operations along with the League of Nations in 1946, but UNESCO would continue and expand activities begun by the ICIC after the Second World War.[9]

The ICIC may be seen as an early-twentieth-century example of science diplomacy. During 2022, the centenary of its creation resulted in various conferences and meetings, allowing historians, legal scholars and international relations experts to examine its achievements but also its limitations. Among these latter is the Committee's lack of diversity and cultural representativeness.[10] As Naomi Oreskes and other scholars have pointed out, diversity is important because the world of science, as indeed the democracies of which science forms an important part, is better off when being confronted with and having to relate to diverse experiences and worldviews.[11] Science works by first testing and then either incorporating or refuting such theories and views. 'A community with diverse values is more likely to identify and challenge prejudicial beliefs embedded in, or masquerading as, scientific theory,' Oreskes argues.[12]

A stronger consensus is reached in the end when different lines of inquiry are raised and debated along the way, not just by scholars around the world, male as well as female, but also by minority and indigenous groups. Precisely because scientific knowledge production is social in character and consensus is arrived at by critical evaluation of diverse knowledge claims, Oreskes maintains, the public may have confidence in the scientific process. Reaching scientific consensus based on arguing and testing a multitude of arguments means, she writes, that:

> we are justified in accepting the results of scientific analysis by scientists as likely to be warranted . . . The critical scrutiny of scientific claims is . . .

done . . . in communities of highly trained, credentialed experts, and through dedicated institutions such as peer-reviewed professional journals, specialist workshops, the annual meetings of scientific societies, and scientific assessments for policy purposes.[13]

When it comes to the public's trust in science, it may at times depend on the perceived impact that scientific results and arguments have on people's lives, Oreskes furthermore suggests. For some people, the issue is less a distrust of science than a fear of what science (and scientists) may do to their way of life and their values and beliefs. A deeply religious person may see what scientists themselves consider great progress within, say, stem cell research or the use of the genome editing tool known as CRISPR/Cas9[14], as leading to a direct repudiation of their religious views and to everything they hold most dear. Likewise, someone who very much believes that power should be close to the people and that the control a state or federal government exerts over the lives of its citizens should be as minimal as possible, would be inclined to fight public vaccine or face mask mandates.[15]

In a somewhat different context, Justice Elena Kagan has also given considerable attention to the issue of public trust. Her focus is on the loss of public trust in the US Supreme Court to which she was herself appointed by President Barack Obama in 2009.[16] The US Supreme Court obviously offers a case study in public trust that is different from that of academia in various respects. We bring up Justice Kagan here, though, because we think we can learn a thing or two from her about how to enhance public trust by disseminating our research in such a way that it becomes immediately understandable to everyone. Argumentation, or the art of arguing, involves dialogue; scientific freedom means freedom to test and to argue, but not to force one's results on others, as a colleague of ours at the University of Copenhagen recently put it.[17]

It is especially in her dissenting opinions that Kagan has endeavored to explain her reasoning and that of her colleagues as a way of promoting a better understanding of how the judiciary works in practice. She published her first dissent already in her first term as a justice on the Court in 2010. But it is especially during the past few years, as the Court has become increasingly conservative, that she has used her dissenting opinions to present legal complexities 'without legalese, in a form of translation,' writes Lincoln Caplan in a 2022 portrait of Kagan. He continues: 'Most outstanding writing by justices still needs unpacking, with definitions of terms, explanations of precedents, and the like. She [does] the unpacking. Her writing [sounds] like she [is] talking, in plain terms and relatively short sentences that a non-lawyer [can] grasp.'[18]

Kagan's dissent in the important case of *West Virginia v. Environmental Protection Agency*, which was decided in June 2022, is a good example—not least in our context, as it concerns Supreme Court decisions on scientific issues. The majority decision, written by Chief Justice John Roberts, sided with Republican-led states and coal companies that the Clean Air Act from 1970 does not provide the Environmental Protection Agency (EPA) with expansive power over carbon emissions from power plants.[19] In her dissent, Kagan explained that Congress and everyone else need and count on the expertise and experience to be found in agencies such as the EPA. This is especially the case when there is a scientific or technical dimension involved, climate change being a case in point. 'Whatever else this Court may know about,' she concluded, 'it does not have a clue about how to address climate change. And let's say the obvious: The stakes here are high. Yet the Court today prevents congressionally authorized agency action to curb power plants' carbon dioxide emissions. The Court appoints itself—instead of Congress or the expert agency—the decisionmaker on climate policy. I cannot think of many things more frightening.'[20]

Elena Kagan is clearly a gifted writer. But, perhaps just as importantly, she does not lecture to or talk down to the public; instead, she addresses them as fellow citizens and stakeholders in American democracy.[21] This is crucial in a political environment such as the American one that is divided to the point of being toxic—and where the word 'elitist' does not warrant respect, but seems instead to have entered the culture wars as a populist signal of criticism against those who think that they may know better than everyone else. How to avoid that kind of hated 'elitism' is the task scholars currently face, not just in the United States but around the world, when they attempt to engage with the public and to honor Article 15(2) and its demand for scientific dissemination as a part of the right to science.

When it comes to the public's engagement with and confidence in science, the visibility of research outputs seems to play a large role: with increased scientific visibility comes increased public interest. Therefore, write Esther Marin-Gonzales et al., 'dissemination and communication of research should be considered as an integral part of any research project.'[22] A level of curation should be built into the research process from the very beginning just as the translation of data and results into a language that the public can relate to ought to be a part of any funding scheme.[23]

CREATING TRUST THROUGH
BETTER COMMUNICATION

Scientists and scholars are well aware that 'effective dissemination and communication [of their research output] are vital to ensure that the conducted research has a social, political, or economical impact.'[24] They furthermore realise, as surveys of what motivates scientists to disseminate their research show, that visibility in the media may also benefit them at a more personal level.[25] From what Beck et al. call a 'value capture perspective,' in addition to the monetary rewards that may be involved in the shape of patenting, licensing or consultancy fees, public dissemination of their research may also lead to an increase of social recognition and reputation: 'These outcomes are considered valuable due to scientists' individual needs, such as the struggle for academic survival (e.g., position), ego-identity needs (e.g., social desirability), as well as the desire to make a societal impact.'[26]

Most scientists consider responding to journalists a professional duty.[27] Entering into dialogue with the public is a challenge, however, as scientists have typically had no formal training in science communication. Many have watched colleagues getting into trouble when attempting to share their research openly—especially when public venues such as social media are involved. In these media, it is easy to twist arguments, willfully misunderstand, and turn against people—and it can be painfully exhausting to try to point out misinterpretations and defend arguments. Attempting to remain as objective and fair as possible for a scholar means offering arguments of the 'on the one hand—but then again, on the other hand'–type. But there is only rarely time and opportunity to make such elaborate arguments. Most often, what is asked of a scholar trying to disseminate their research is to condense sometimes quite intricate results into a couple of sentences, giving the public the impression that they are simplifying and taking sides.

Aware of their own inadequacy as communicators, some scientists seek the help of professional science journalists. This is an obvious solution, as these journalists are trained communicators and may help bridge the gap between scientists, policymakers, and the public. As science journalists attempt to avoid jargon use and increase straightforwardness in science communication, however, accuracy and important details may get lost in the process, just as inaccurate arguments may not be countered directly and on the spot. Likewise, science journalists may not necessarily prioritise—or get across to the public—the excitement entailed in the creation and provision of knowledge, which is the core function of science.

Could citizen scientists perhaps help? Citizen science is the involvement of the public in scientific research. Nonscientists can participate in the process of gathering data and of mapping natural sites, for example, or they can engage in decision-making about policy issues involving scientific or technical elements. For those interested in citizen or Do-it-Yourself (DiY) science, the activities of community-based science research hubs combine a desire to learn with a wish to engage with the possibilities that science and technology offer to the development of local communities and society in general. Typically referring to the concept of 'open science,' the proponents of citizen and DiY science advocate for the democratisation and sharing of scientific data which they argue may lead to a much-needed demystification of science.[28]

As beneficent as opening up the processes of science and technology to the public may be, however, citizen and DiY science do encompass multiple—at times contradictory—understandings and expectations concerning the scientific process and, just as importantly, the possible ethical issues implicated in this process. This may end up posing risks not only to the practitioners involved, but also to human health and the environment.[29] One example is the use of DNA technology to test for genetic predispositions for disease. This can be done commercially, but also in DiY biology labs that explain to non-scientists how to use this technology. The enthusiasm for self-administered genetic testing raises ethical questions such as, 'What if someone discovers something disturbing about their own genome? Does anyone bear responsibility here for mitigating risk or harm, or how is this best achieved'?[30] While most tend to agree that citizen and DiY science are positive phenomena from which the world of science will benefit, there are clearly issues involved here that need to be debated. This is reflected in the divergent literature which, as David Sarpong et al. sum it up, 'contains many examples on the one hand of how DiY science could serve as a potential means to pushing scientific frontiers, but on the other hand highlighting the perils of hacking or underlying security and ethical implications of DiY practices.'[31]

In terms of the engagement of the public with policymaking involving scientific issues, participation must involve some degree of decision-making. From a human rights perspective, as first UN Special Rapporteur in the field of cultural rights Farida Shaheed noted in her 2012 report on the right to science, the cultural human rights to science and culture, which are mentioned side-by-side in both Article 27 of the UDHR and Article 15 of the ICESCR, should be read together. In addition, they should also be read "in conjunction with, in particular the right of all peoples to self-determination and the rights of everyone to take part in the conduct of public affairs."[32] Public consultations on scientific advances and their implications should be organised so as to protect everyone, but especially marginalised persons and groups, from dual-use science and technology.[33] Likewise, the participation of individuals and communities in decisions concerning research priorities and funding is necessary for the promotion of appropriate research that can address societal needs.[34] What this means is that when we view the world of science through a human rights lens, scientific research and scientific freedom cannot be completely free, but must be conducted in a socially and ethically responsible manner.[35]

Does the kind of public participation in science policymaking called for here mean, to take a concrete example, that in modern democracies the public should be able to decide what sort of research we and our university colleagues undertake? In the context of one of our home countries, Denmark, for example, we are publicly funded and according to the Danish law on universities, as central conveyors of knowledge and culture Danish universities 'must exchange knowledge and competences with Danish society and encourage their employees to participate in public debate.'[36] In practice, that knowledge-exchange will be carried out, not by the universities themselves, but by individual scholars such as us. As we have argued elsewhere, we see this situation as a potential problem for scientific freedom—our own as well as that of our colleagues.[37] If the public's participation in decision-making concerning science results in funding being given only to topics defined by politicians in response to the perceived needs of the general public (or lobby groups, for that matter), this may lead to less funding for basic research and more funding being allocated to strategic or applied research.

Unlike strategic or applied research, basic or fundamental research is curiosity-driven, scholars being motivated by a wish to expand human knowledge without having specific applications or commercial objectives in mind.[38] Somewhat counterintuitively, experience shows that it is often curiosity-driven research which later becomes the basis for solving real-world problems. Many of the key advancements of modern science are based on knowledge gained in basic science by people who were uncertain of where their research was leading and who worked without a clear time frame. One example is the development of computer chips, which would not have been possible without quantum mechanics; another is GPS devices that build on Einstein's theory of relativity.

The problem is that there is no clear way to predict which basic scientific research will lead to major discoveries. Currently, in many Western democracies, public funding bodies seem to be increasingly interested in applied research that can promise practical outcomes and products of immediate value:

> The orientation towards applicability, problem solving, and impact has become explicit in national research projects funded by governments and in the evaluation criteria used by universities. This has led to concern that the focus on applicability in funding and evaluation has created extrinsic incentives and

constraints on academic research and has encouraged a shift from basic to applied research in many North American and European countries.[39]

The recent move of orienting public research towards societal challenges through 'missions' adds to this orientation toward applicability. Following recent European Union mission-driven policies that address grand societal challenges and the UN Sustainable Development Goals, Denmark in 2020 introduced a mission-driven research strategy, for example, that specifically targets carbon capture, green fuels, climate-friendly agriculture and food production, and the re-use and reduction of plastic waste.[40] The Danish government now aims to facilitate collaboration among central actors of the green transition such as research funding organisations, universities, and also the private sector. The idea is not just to encourage a more coherent effort across the whole value chain from basic research to commercialisation, but also to engage more actively with societal challenges, local as well as global.[41]

The result is, write Kaare Aagaard et al., a much stronger 'state-led directionality' and coordination which is 'likely to have implications for the criteria used to assess and select projects for funding. For instance, instruments that deploy mission-driven policies would likely include an assessment of the ability of proposed projects to contribute to solutions to a given societal challenge or mission, in addition to conventional assessment.'[42] This raises a number of questions; how may, the authors ask, the stronger role politicians want to have in directing research to particular focus areas be balanced with the need for high-quality academic research, for example?[43]

And what will be the impact on scientific freedom? As achieving funding for basic research becomes increasingly difficult, some scholars may feel forced to alter their 'blue-skies' research ambitions and to target their research ideas specifically to concrete strategic or mission-driven research funding calls—experiencing this move as a weakening, if not downright denial, of their scientific freedom. These scholars may well see the increasingly active participation of individuals, communities, and politicians in setting research priorities as contributing to this weakening of their scientific freedom.[44] As we noted in Chapter Two, mission-driven research is likely to involve a slowing of the rate of general scientific progress and should therefore only be undertaken to address the most pressing needs and allow for as much freedom in the choice of methods and approaches as possible.

Beyond the issue of funding itself, citizen participation in science also has relevance for discussions concerning science dissemination and Article 15(2) ICESCR. As Hans Peter Peters argues, for example, citizen science and an increased public participation in science have led to a change in emphasis from 'the public understanding of science' to the 'public engagement with science and technology.'[45] In an attempt to find a solution to the crisis of trust in science, the focus amongst science communicators and journalists is no longer so much on the communication of knowledge, but rather on improving the relationship between science, scientists and the public. Providing entertainment and good experiences for the public has become a goal in itself for these communicators who seem to forget, writes Peters, that 'plain dissemination of knowledge remains an important function of science communication.'[46] In and of itself, the inclusion of non-scientists in the construction of knowledge is not a bad idea, he concludes; it is just that when it comes to the dissemination of this knowledge, we should not forget that 'the core function of science is creation and provision of knowledge and hence this should be prioritized in the public communication of science.'[47]

We tend to agree with Peters. When less attention is paid to the actual communication of scholarly knowledge, the pursuit of knowledge and truth that the framers of the UDHR and the ICESCR saw as the main objective of the right to science recedes into the background.[48] Scientists therefore need to be actively involved themselves in communicating to the public what their science is about, what its potential is, and why science is important as a public good. This in no way eliminates the need for professional science journalists. To honor the promise of Article 15(2) and its importance for the right to science, scientists would still very much benefit from being coached by professional science journalists who can show them how to engage with the public in a respectful, non-paternalistic/elitist way. At a time when the public becomes more and more involved in science-policy decision-making, communication skills need, we would furthermore argue, to be included as a part of the educational curriculum in STEM and other university disciplines.[49]

Becoming more visible and reaching the wider public gives scientists the chance to show the importance of their line of research, thereby increasing their chances of securing future funding. Acquiring the skills that are necessary to grasp the attention of the public and their political representatives can make all the difference to the individual scholar and their professional career prospects. To produce the kind of groundbreaking and excellent research, basic or applied, on which that dissemination to the public must necessarily be based, individual scientists need access to scholarly results achieved by colleagues around the world. This is where the obligation of States to recognise the importance of international contacts and co-operation in the scientific field, outlined in Article 15(4) ICESCR, becomes relevant. In the next section, we will look at the fourth part of Article 15 as it relates to scientific freedom.

SCIENTIFIC FREEDOM AND
INTERNATIONAL COOPERATION

As noted above, the creation of the International Committee on Intellectual Cooperation (ICIC) may be seen as an early-twentieth-century example of science diplomacy. The remit and work of the ICIC being passed on to UNESCO, moreover, the Committee is also the precursor of the UNESCO Chairs and UNITWIN network to which one of the authors' own UNESCO Chair in Cultural Rights belongs. The UNESCO Chairs/UNITWIN Program was launched in 1992 and today involves more than 850 different institutions in 117 countries.[50] The idea behind it was to promote international inter-university cooperation and knowledge sharing in the key UNESCO priority areas—education, sciences, culture, and communication and information—so as 'to accelerate mutual understanding and a more perfect knowledge of each other's lives.'[51] This promotion of knowledge sharing across the world echoes the aim of Article 15(4), namely to 'recognize the benefits to be derived from the encouragement and development of international contacts and co-operation in the scientific and cultural fields.'

UNESCO's remit spans the four core cultural rights: the rights to education, culture and science, and authors' rights. The three latter rights are mentioned side-by-side in Article 15,1(a–c) ICESCR; yet, together with related issues concerning intellectual property (IP), authors' rights have mainly been dealt with by the international intergovernmental organization, World Intellectual Property Organisation (WIPO). Like the World Trade Organisation (WTO), with which it has a mutually supportive relationship, WIPO is deeply rooted in the multilateral trading system. This is reflected in the mandate that WIPO has when it comes to standard-setting for authors' rights—a mandate which is significantly different from that of UNESCO. Whereas WIPO seeks to safeguard authors' rights by means of legal protection through IP law, UNESCO works through the adoption of soft-law measures such as declarations and recommendations to ensure authors' (and creators') "moral and material interests resulting from any scientific, literary or artistic production of which he is the author" (Article 15,1(c)).

The official WIPO position on authors' rights, as explained on the WIPO website, is that

> copyright (or author's right) is a legal term used to describe the rights that creators have over their literary and artistic works . . . There are two types of rights under copyright: economic rights, which allow the rights owner to derive financial reward from the use of his works by others, and moral rights, which protect the non-economic interests of the author.[52]

By contrast, the position adopted by several human rights scholars is that authors' rights are different from copyright and patents and that unlike authors' rights, IP rights are not human rights. Whereas authors' rights are personal (or inalienable, non-transferable) rights that concern the reputation and integrity of creators (the rights of attribution and of objecting to any mutilation or other distortion of one's work), IP is a monopoly that someone has for a limited period of time only. After this time—which is typically twenty years for patents—the IP-protected item or knowledge falls into the public domain and can be used free of charge by others. IP rights are not personal, but alienable and can be given away to others. The IP regime is, writes Farida Shaheed, 'a temporary monopoly,' and "the rights of authors protected by human rights instruments are not to be equated with 'intellectual property rights'."[53]

Along with all other human rights, cultural human rights such as authors' rights belong within the public goods–oriented system of UNESCO (and other UN institutions), whereas IP rights are private rights that are regulated under the auspices of the WIPO/WTO system of privatised property rights. The two systems have evolved separately, and they work in different ways.[54] Somewhat confusedly, though, the international human rights system is not completely consistent. Whereas most human rights instruments talk about authors' rights, IP rights are mentioned directly in Article 31(1) of the 2007 United Nations Declaration of the Rights of Indigenous Peoples (UNDRIP): '[Indigenous peoples] have the right to maintain, control, protect and develop their intellectual property over such cultural heritage, traditional knowledge, and traditional cultural expressions.'[55] Likewise, Article 17(2) of the European Charter of Fundamental Rights says that 'intellectual property shall be protected.'[56]

The question of ownership of knowledge has a direct bearing on access to knowledge and therefore also on the issue of scientific freedom. For scientists themselves as well as for the public, having a right to science means having a right to access the latest scientific discoveries made by their national as well as their international colleagues. In a report from 2013, prepared for the American Association for the Advancement of Science (AAAS), Margaret Vitullo and Jessica Wyndham surveyed the perspectives of American scientists on the right to science.[57] Based on analyses of focus groups conducted with 145 American scientists from various disciplines, they identified various types of access and depicted these on 'a fluid and bi-directional continuum' stretching from the interests of the public ('access for general public') to those of scientists ('access for scientists').[58]

This continuum of access can be obstructed in various ways. By privatising and taking out of the public domain research results, IP may constitute one such obstacle. Many journals and other research outlet venues are now behind paywalls, and this makes it not just very difficult, but also very expensive for both scientists and the general public to get access. Christine Mitchell describes the situation in this way:

> Much science is done in academic and government settings or for-profit companies. Scientists who share their findings in scholarly articles, monographs, and books for the academic market are generally not paid to publish, and their data, finding, and publications are usually not freely available to the general public. Instead, scientists submit their work to journals and publishing corporations, such as Elsevier, who handle scientific peer review and publication, and the information is privatized and put behind a paywall. Libraries then pay an institutional fee for faculty to have access, or individuals may pay a fee per article. Sometimes authors pay publishing companies a fee as well, usually in the thousands of dollars, for their work to be freely available—typically labelled 'open access' – in which case the for-profit publishing company collects fees from both the universities and authors.[59]

The international geo-political scene constitutes a second potential obstacle to access. Again, in a global world challenged by problems that can only be solved globally, when scientists do not have access to global scientific results, their academic freedom suffers. This and other issues at the intersection of science, policy, and international relations have long been debated within the discipline of science diplomacy.[60] Depending on the definitions used, science diplomacy dates to at least the mid-nineteenth century, or to early modernity. As an academic discipline, though, it is a relatively recent field occupied by science advisors, diplomats and scholars.[61] Although typically serving the purpose of advancing specific national power interests, science diplomacy has also aimed to address common problems and to build constructive international partnerships. One famous example is the European research center CERN, one of whose co-founders was Danish Nobel Prize winner Niels Bohr.

Having been involved with the Manhattan Project, Bohr realised that the traditional tools and tactics of foreign policy were out of step with the complexity of modern science and technology. During WWII, he warned decision-makers that if left to national competition, the nuclear revolution would lead to destructive political developments.[62] In June 1950, Bohr sent an Open Letter to the United Nations in which he argued that the only way to curtail this danger and to maximise benefits from recent scientific breakthroughs was to share science and technology across geographical and academic borders.[63]

With the bitter trade battle between the United States and China and the recent outbreak of war in Ukraine, Bohr's ideas are as timely as ever. As a prerequisite for peace, openness must, as he argued, be encouraged, and scientists have an important role to play in this. In the twenty-first century, many of the defining challenges—climate change, poverty reduction, global pandemics such as COVID-19 and nuclear disarmament—have scientific dimensions. Realising that no one country will be able to meet these challenges on its own, the EU wants to play an active role in international science diplomacy.[64] The German government also recently expressed a wish to turn the Federal Foreign Office into a laboratory for science diplomacy.[65]

UN instruments recognise the important role of scientists in helping to lay the foundation for a peaceful, sustainable world by collaborating broadly with citizen scientists and colleagues in other fields; by engaging with legislators and policymakers to help base policies on evidence; and by paying due regard to the ethical implications of their research.[66] As Bohr and other scientists showed with regard to the Cold War, the best prerequisite for overcoming political mistrust and for working towards common, evidence-based solutions is a sense of shared urgency to solve critical global problems.

The centrality and importance of access, in the current context especially to data, for meaningful implementation of Article 15 was acknowledged by all AAAS focus group participants. When national security is at stake, dual-use science and technology are involved, or a need for consideration of privacy and confidentiality arises, providing such access becomes difficult. Yet, participants still thought that 'the default position should be for data to be openly available and accessible.'[67] It was important for them as researchers that the collection of data should not be guided by political considerations, but instead by peer review and scientific standards and that it be made accessible for ongoing research. Beyond increased capacity for the storing of data, what access requires, they agreed, are 'international agreements on how to store data so it can easily be accessed, and wide dissemination—to scientists, policy makers, and to the public, both nationally and internationally.'[68]

The US-China trade war has made a wide dissemination of research results increasingly difficult. Just as science and technology played a major role during the Cold War, so they are at the heart of the current tension in trade relations between the two countries: "The vast majority of the issue is tech, not trade . . . The 1960s brought us the global space race. We are now in the next state of the global data race . . . Those countries that are able to harness data from both an offensive and a defensive perspective for economic prosperity and national security, will be ahead of others.'[69] In this science and technology competition, scientific research has, as Andrew Silver puts it, 'become collateral damage.'[70]

CONCLUDING REMARKS

A clear benefit of scientific progress is that scientific knowledge is used in decision-making and policies, which should, as far as possible, be based on the best available scientific evidence. States should endeavour to align their policies with the best scientific evidence available. They should, furthermore, promote public trust and support for sciences throughout society and a culture of active citizen engagement with science, particularly through a vigorous and informed democratic debate on the production and use of scientific knowledge, and a dialogue between the scientific community and society.[71]

One of the 'special topics of broad application,' mentioned in the General Comment No. 25 on the right to science, concerns the role of science-based policymaking. The CESCR does not say in so many words that the right to science encompasses the right to base public policies on the best available scientific evidence—but it comes close! In the same paragraph in which this is mentioned, the Committee also underscores the importance of dialogue between the world of science and the public.

Both evidence-based policymaking and public dialogues presuppose that scientific knowledge is generated. Without scientific results there will be nothing to base public policy on, just as there will be nothing to share and debate with the public. This sounds very basic and easy to grasp—but what is sometimes forgotten is that the foundation for all scientific knowledge creation is scientific freedom, and that scientific freedom cannot exist without global access to scientific results. Such access is currently jeopardised by geo-political events such as the trade wars between the United States and China and the Russian invasion, more recently, of Ukraine. National security concerns, but also—at least with respect to the trade wars—issues concerning IP are involved. But then, as Clare Wells wrote a while back in an article on UNESCO's recommendations on science and culture, 'although scientists may be seen to derive power from their knowledge and skills, the knowledge in question is of a type (militarily and/or commercially relevant) whose use States in general and powerful employers have an interest in controlling.'[72]

Given the pervasive, global impact of modern science and technology, increasingly larger numbers of scientists today dabble in some use and exchange of science and technology, involving the public and their elected political representatives in addition to their fellow-scientists around the world. This exchange starts with access to global research results and when such access is obstructed, scientific freedom suffers in the process. Without scientific freedom, there will not only be no creation of scientific knowledge;

there will also be no joint efforts to address global challenges. This leaves very little space for the recognition of 'the benefits to be derived from the encouragement and development of international contacts and co-operation in the scientific and cultural fields.'

PART II

SAFIRES: A Conceptual Model of Scientific Freedom Obligations under International Human Rights Law

Chapter 5

The SAFIRES (Scientific and Academic Freedom as Integral Elements of the Right to Enjoy the Benefits of Science) Model

As we have seen, the drafters of the international bill of rights viewed scientific and academic freedom as worthy of protection under international human rights law because they contribute towards the production of knowledge and technology, and towards the discussion and dissemination of ideas. Although scientific and academic freedoms are, to varying extents, protected by national constitutions and other sources of domestic law, their inclusion in international human rights law is significant because it places a set of separate and additional obligations on States parties to the ICESCR. Their inclusion is also significant because it means that the ICESCR treaty framework, as well as rules and interpretive tools applicable to international human rights law more generally, can be applied to develop a more detailed understanding of the obligations that States must guarantee and facilitate these freedoms.

Some of these rules are contained in the ICESCR itself. Articles 2 and 4 ICESCR, for example, contain rules for the implementation and limitation of economic, social, and cultural rights. Others derive from the specific context of international human rights law, such as interpretive devices developed by treaty bodies tasked with monitoring the implementation of international human rights law treaties. Because the ICESCR is a treaty, more general customary rules for treaty interpretation as laid out in the 1969 Vienna Convention on the Law of Treaties are also relevant.[1]

In previous works, we have drawn on some of these tools to develop a Four Step Test for the evaluation of policy compliance with right to science obligations.[2] Our aim was to describe how uncontroversial and widely accepted interpretive devices can be applied to the right to science to make it easier to understand and work with this right in practice. In this chapter, we introduce the SAFIRES model (Scientific and Academic Freedom as Integral elements of the Right to Enjoy the benefits of Science), which is meant to facilitate the practical application of right to science-based scientific freedom standards to policymaking, evaluation and analysis.

The chapter is divided into three parts. The first part looks at the ICESCR treaty framework as well as State obligations and limitation criteria under the ICESCR. The second part starts with a brief outline of the Four Step Test as well as of the rules of interpretation on which it is based. We then use these to extend and adapt the test into a conceptual model of scientific freedom obligations under international human rights law: the SAFIRES model. In the third and final part of the chapter, we introduce three international instruments of great relevance to scientific freedom as well as science and society more broadly. These are the 2017 UNESCO Recommendation on Science and Scientific Researchers; the 2020 General Comment; and the 2021 UNESCO Recommendation on Open Science. The following chapter applies the model to these instruments, aiming to provide practical examples of the usefulness of the understanding of scientific freedom advanced here. We furthermore wish to draw attention to, and to critique, those provisions in these instruments which appear to us suspect from the point of view of scientific freedom as a constitutive element of the right to science.

Our hope is that those tasked with understanding and implementing human rights obligations, as well as those laboring to hold governments to account for these, will find the SAFIRES model a useful tool in litigating, advocating, legislating and otherwise affecting science-related policy.

THE ICESCR TREATY FRAMEWORK, STATE OBLIGATIONS AND LIMITATIONS CRITERIA UNDER THE ICESCR

By signing and ratifying the ICESCR, States parties voluntarily assume the obligations codified in that treaty. Articles 16 through 22 of the ICESCR grant various monitoring powers to the UN Economic and Social Council (ECOSOC) and require States parties to provide reports detailing their progress towards implementing Covenant rights. In 1986, ECOSOC delegated its monitoring function and powers to the Committee on Economic, Social and Cultural Rights (CESCR), which is composed of eighteen independent experts.[3]

Based on the expertise of its members, on the reports it receives from States parties on their implementation of Covenant rights, as well as on consultations with States and civil society, the CESCR issues guidance on the interpretation of obligations under the ICESCR meant to facilitate the ongoing process of clarifying and developing human rights norms. Of these, the issuance of General Comments is particularly relevant, as they contain the CESCR's interpretation of the normative content of rights under the ICESCR.[4]

General Comments provide practical guidance to States parties, help structure the CESCR's work, and serve as reference points for evaluating individual and inter-State communications. While the legal status of General Comments is debated, they are widely acknowledged as soft law instruments with presumptive authority.[5] They address necessary questions and derive their legitimacy from the expertise of the CESCR members, the treaty-based mandate of the Committee, and the extent to which judges, States, and advocates rely on their interpretations in practice. States must consider the contents of General Comments in good faith, and any opposing interpretations must contend with the CESCR's interpretations as the presumptively correct starting point.[6] We shall deal in more detail with the importance of General Comments below.

State Obligations under the ICESCR

The ICESCR itself contains some guidance as to how States parties are to implement these obligations. Article 2(1) ICESCR, for example, contains what is known as the rule of progressive realisation of the rights contained in the Covenant:

> Each State Party to the present Covenant undertakes to take steps, individually and through international assistance and co-operation, especially economic and technical, to the maximum of its available resources, with a view to achieving progressively the full realization of the rights recognized in the present Covenant by all appropriate means, including particularly the adoption of legislative measures.

The rule of progressive realisation is meant to recognise that the various obligations in the Covenant must compete for scarce resources, and that it will not be possible to immediately realise the full enjoyment of all the rights contained in the Covenant. Nevertheless, the CESCR has interpreted the rule of progressive realisation to mean that steps must be taken within a 'reasonably short period of time,' and that the obligation 'implies not only legislative measures, but also administrative, financial, educational, social and other measures, including judicial remedies.'[7]

Furthermore, since realisation must be progressive, retrogressive measures which reduce the level of realisation of a Covenant right are generally prohibited.[8] In the context of the right to science, the CESCR lists the following examples of retrogressive measures: '[T]he removal of programmes or policies necessary for the conservation, the development and the diffusion of science; the imposition of barriers to education and information on science; the imposition of barriers to citizen participation in scientific activities, including misinformation intended to erode citizen understanding and respect for science and scientific research; and the adoption of legal and policy changes that reduce the extent of international collaboration on science.'[9]

Some obligations are treated as being of such significance that they must be realised immediately, without consideration of available resources. Article 2(2) of the ICESCR, for example, mandates the enjoyment of the rights outlined in the document in a non-discriminatory manner, and the CESCR has stated that, 'States Parties are under an immediate obligation to eliminate all forms of discrimination against individuals and groups in their enjoyment of ESCRs.'[10]

The CESCR has also developed the concept of a core level of enjoyment of economic, social, and cultural rights, which States parties must provide 'as a matter of priority' or else demonstrate 'that [they have] made every reasonable effort to comply with them.'[11] These core obligations include the removal of 'limitations to the freedom of scientific research that are incompatible with article 4 of the Covenant.'[12] Other core obligations of relevance to our understanding of scientific freedom include the removal of barriers that limit access to science and scientific knowledge; the elimination of barriers to equal participation of women and girls; the development of a participatory framework for the conservation, the development, and the diffusion of science; ensuring access to basic education and skills; promotion of accurate scientific information; protection against pseudoscience; and fostering international cooperation in science.[13]

Both Articles 2(1) and 15(4) of the ICESCR highlight the significance of international cooperation and aid. In the General Comment on Article 2, the Committee clarifies that the term "available resources" should encompass resources obtained through international collaboration. Furthermore, it asserts that wealthier nations bear a unique responsibility to support less affluent countries in fulfilling their obligations under the ICESCR.[14]

Limitations Criteria

The ICESCR treaty framework recognises that the enjoyment of human rights cannot always be guaranteed. States may have to make trade-offs between the promotion of human rights and other functions, as well as between various rights. Circumstances such as national emergencies may furthermore have an impact on their ability to progressively realise the enjoyment of human rights.

In Article 4, the ICESCR lays out detailed rules for when any such limitations may be legitimate:

> The States Parties to the present Covenant recognize that, in the enjoyment of those rights provided by the State in conformity with the present Covenant, the State may subject such rights only to such limitations as are determined by law only in so far as this may be compatible with the nature of these rights and solely for the purpose of promoting the general welfare in a democratic society.

The CESCR highlights three essential elements of ICESCR Article 4 for limitations to be legitimate: they must be determined by law, compatible with the nature of the rights in the Covenant and enacted solely for promoting general welfare in a democratic society.[15]

The first criterion means that any legitimate limitations must observe the rule of law and be issued through proper channels. The Limburg principles outline several criteria for such limitations, including being non-arbitrary, non-discriminatory, clear and accessible to everyone, with safeguards against illegal or abusive imposition.[16]

Compatibility with the nature of rights in the ICESCR entails alignment with fundamental human rights principles, including dignity, the rule of law, equality, non-discrimination and universality. It also means that limitations of a particular right must not nullify the enjoyment of that right. One way of understanding this is to say that any limitation must be compatible with the minimum core obligations of the right limited and be proportionate to the aim pursued.[17] The proportionality requirement involves selecting an appropriate aim, choosing the option least restrictive to economic, social, and cultural rights, and ensuring that the burdens imposed do not outweigh the benefits of the limitation.

The focus on general welfare emphasises inclusiveness and the equal importance of each individual's preferences, interests and well-being. This principle makes it harder to justify limitations that benefit only a select group and encourages policymakers to consider the interests of everyone in society. While some argue that public order, public morality, and national security also serve the general welfare, suggestions to include these aims were rejected during the drafting of the ICESCR. This implies that the drafters had a concept of the general welfare in mind that centers on social and economic well-being and individual happiness.[18]

The reference to a democratic society aims to prevent the abuse of the limitation clause by authoritarian or non-representative governments, which might define general welfare in ways that do not correspond to the aggregate welfare of everyone in society.[19] What matters is a governance structure that values the interests of each member of society equally.

THE FOUR STEP TEST AND SAFIRES

Building on these rules of interpretation, we developed the Four Step Test for policy evaluation based on the right to science.

Four Step Test

The first step of the model involves applying definitions of key terms, such as 'science,' to determine whether an issue or policy falls under the scope of the right to science. In the second step, issues and policies are sorted according to whether they help or hinder the realisation of the right to science, and if the latter, whether they are best classified as derogations, retrogressive measures, or limitations. In the third step, policies that limit enjoyment of the right are assessed according to the criteria for legitimate limitation laid out in Article 4 ICESCR. The fourth and final step is to select the optimal policy solution from options that have made it through the previous three steps.

A policy option that passes the Four Step Test meets these criteria: it maximises general welfare in a democratic society in a way that respects the rule of law, is based on science, is compatible with the nature of economic, social, and cultural rights, respects fundamental human rights principles, and balances its impact on all human rights. We argue that such outcomes are likely superior to the criteria used in current science policy decision-making, as they prioritise both general and individual interests.

SAFIRES

Zooming in on the scientific freedom part of the right to science, the Scientific and Academic Freedom as Integral elements of the Right to Enjoy the benefits of Science model adjusts the Four Step Test into a conceptual model of scientific freedom obligations under international human rights law. SAFIRES consists of three sequential stages. First, whether the right to science is applicable to the issue or policy is assessed. Second, the likely impacts of the policy are examined against minimum human rights standards. Finally, ways to improve the policy from a human rights perspective are considered.

The issues to be raised and the questions to be asked in each of these three stages of the SAFIRES model are these:

I. *Scope determination*
   a. Does the issue or policy involve the government or purely the private sector?

b. Does the issue or policy fit within the scope of science and scientific freedom?

c. Does the issue or policy affect academics, defined as people researching at a university or government research institute?

II. *Minimum standards*

a. Does the issue or policy meet the standard of core obligations?

b. Does it restrict the freedom to choose research topics, to participate freely in scientific research, or freely to access the information and resources necessary to carry out research?

c. If yes, does it meet Article 4 limitations criteria?

III. *Optimal choice*

a. Has the policy been developed with representation and participation of those affected?

b. Has the policy been developed using the best available scientific evidence?

c. Have its benefits been maximised and costs, harms, and risks minimised?

Let us look at each of these three stages in a little more detail.

Scope Determination

The first, preliminary stage involves ascertaining whether scientific freedom obligations under the right to science arise in the first place. For this to be the case, a policy or issue must involve the public sector, as human rights obligations derive from treaties which bind only governments. Although there are human rights standards specifically developed for business, the private sector is not legally bound to observe these or human rights more generally in the same way that governments are. For these reasons, the private sector is not of main concern to our analysis.

Second, scientific freedom under the right to science only applies to areas that fall within their scope. A second preliminary step therefore involves comparing the policy area to our definitions of 'science' and 'scientific freedom.' As previously stated, our definition of scientific freedom for the purposes of this book includes the right to pursue scientific research without undue external influence, to participate in scientific activities, and to access scientific knowledge and resources. This includes freedom from interference (negative freedom) and the availability of resources and support (positive freedom) and extends to all branches of the academy, not just the natural or social sciences.

Where the issue or policy affects academics, defined as people carrying out research at universities and government research institutes, the higher standard of academic freedom is engaged in addition to scientific freedom.

As understood here, though academic freedom overlaps significantly with scientific freedom, it involves a higher level of protection just as it includes protections in areas of relevance primarily to academics, such as tenure.

Minimum Standards

Once scientific freedom under the right to science has been deemed applicable to a policy or issue, the next stage involves comparing its likely effects to minimum standards for the enjoyment of scientific freedom.

Some of these minimum standards are provided by the CESCR in its list of core obligations.[20]

According to these, policies and laws must neither undermine women's and girls' participation in science and technology, nor contribute to the spread of mis- and disinformation. Likewise, the policies and laws in question cannot limit international cooperation and people's access to scientific knowledge or its applications beyond what is acceptable according to Article 4 ICESCR. Policies must also provide, at a minimum, access to basic science education in line with the best available scientific knowledge. Any policies falling below these standards constitute a violation of the right to science.

Policies must, in addition, not restrict the freedom to choose research topics, to participate freely in scientific research, or freely to access the information and resources necessary to carry out research unless they conform to the Article 4 limitations criteria.

Optimal Choice

A policy that clears both previous stages is compatible with the scientific freedom obligations imposed by the right to science. However, it may still be possible to improve a policy from a human rights perspective or to choose the best policy option among several variations.

One way this can be done is to ensure that the policy is developed with the participation of those that are affected. In the case of scientific freedom, the experience and perspectives of both citizen scientists and academics are relevant, as are those of individuals and groups likely to be affected by scientific progress and its applications. Given the nature of modern science and technology, this latter group is large and essentially corresponds to the general public. As is the case with all policy, adequate representation, consent and involvement of stakeholders may lead to the identification of missed errors or to other improvements. This kind of involvement of the public is furthermore crucial regarding anticipating and subsequently avoiding the dangers of dual-use science and technology.

Given the subject matter, policies that have an impact on science and scientific freedom should themselves be informed by scientific evidence. The social sciences are an important source of such evidence and must therefore also be kept free from political or other interference (thus facilitating the rest of science). Policies that are based on clearly incorrect or highly questionable scientific assumptions are not compatible with the nature of the right to science and should therefore not be implemented. One example is the ban on research into psychoactive substances on historical and political grounds, which has substantially inhibited promising lines of medical research.[21] Identification of the best available science for policy responses is also a function of seeking the input and participation of scientists and academics in policymaking, for example via science advisory boards or national academies of science. This input and participation should be sought from both social scientists with expertise in policy and scientists with domain expertise in the subject matter impacted by potential policies.

Finally, there are general ethical and moral reasons to attempt to minimise costs, risks, and harms associated with a policy as well as to maximise its benefits. Where a variation on a policy can reach the same objectives quicker, more effectively, or at less cost, that policy should be preferred even if less effective alternatives are also human rights compliant.

APPLYING THE SAFIRES MODEL: A BRIEF INTRODUCTION OF THREE RELEVANT SOFT LAW INSTRUMENTS

In the next chapter, we will use the SAFIRES model to assess three international instruments of great relevance to scientific freedom as well as to science and society more broadly: the 2020 CESCR General Comment as well as the two UNESCO Recommendations on Science and Scientific Researchers (2017) and on Open Science (2021), respectively. The rest of the present chapter is dedicated to a brief presentation of these three soft law instruments.

General Comment No. 25

As previously touched upon, General Comments are interpretative guidance documents issued by treaty bodies to help clarify and develop human rights norms and obligations. While not legally binding, these instruments carry significant persuasive weight due to the expertise, independence and authority of the issuing treaty bodies. General Comments serve various functions, including providing practical guidance to States parties in fulfilling their reporting obligations and offering reference points for evaluating individual and inter-State communications. As interpretations of human rights obligations, General Comments are considered presumptively correct and have, to varying degrees, been relied upon by domestic courts, other treaty bodies and other stakeholders.[22]

In the context of scientific and academic freedom, the most relevant General Comments are No. 25 on Science and Economic, Social and Cultural Rights, published in 2020, and No. 13 on the Right to Education, published in 1999, respectively. Given our focus on scientific freedom as a constitutive element of the right to science, we center our attention here on the former. An excellent discussion of academic freedom under the human right to education can be found in the footnote associated with this sentence.[23]

Given the lack of judicial pronouncements on, and the general neglect of, the right to science, General Comment No. 25 has particular significance for the field and was long awaited. Its development is detailed by Mikel Mancisidor, one of the principal authors of the document, in his chapter in Part Three of this book. The document begins by defining the normative content of the right, such as access, enjoyment of benefits, participation and freedom to do science. It then analyses the right in accordance with widely used frameworks for understanding human rights obligations, such as the AAAQ (availability, accessibility, acceptability, and quality) and the tripartite typology of obligations (respect, protect, fulfill). Finally, General Comment No. 25 emphasises the importance of international cooperation in realising the right to science, identifies areas of special application, and discusses domestic implementation of the right, including core obligations.

The General Comment's understanding of scientific freedom is, overall, very much in line with that developed here. It makes a clear connection between scientific freedom and scientific progress: 'In order to flourish and develop, science requires the robust protection of freedom of research.'[24] In addition to the protection from interference,

> States parties [. . .] also have a positive duty to actively promote the advancement of science through, inter alia, education and investment in science and technology. This includes approving policies and regulations that foster scientific research, allocating appropriate resources in budgets and generally creating an enabling and participatory environment for the conservation, the development and the diffusion of science and technology. This implies, inter alia, protection and promotion of academic and scientific freedom, including freedom of expression and freedom to seek, receive and impart scientific information, freedom of association and freedom of movement; guarantees of equal access and participation of all public and private actors; and capacity-building and education.[25]

The General Comment goes on to mention several specific freedoms and entitlements that fall under our elements of freedom to access and freedom to participate in science. Among these are: 1) 'a strong research infrastructure with adequate resources, and adequate financing of scientific education';[26] 2) 'encouraging the widest participation in scientific progress';[27] 3) 'ensuring that private investment in scientific institutions is not used to unduly influence the orientation of research or to restrict the scientific freedom of researchers';[28] and 4) 'adopt[ing] legislative, administrative, budgetary and other measures [. . .] include[ing] education policies, grants, participation tools, dissemination, providing access to the Internet and other sources of knowledge, participation in international cooperation programmes and ensuring appropriate financing.'[29]

In line with our analysis of scientific freedom, the General Comment likewise extends this protection to all individuals engaged in scientific pursuits, regardless of whether these are based at a research institution.[30] Nevertheless, there are a few areas of divergence between the understanding of scientific freedom underlying the SAFIRES model, and that described in the General Comment. These include the definition and scope of 'science' and thus 'scientific freedom'; intellectual property rights; the role of metrics and public accountability in science; the assignment of aims and goals for science; and popular participation in scientific decision-making. These areas of divergence are explored in the following chapter.

UNESCO Standard-Setting Instruments

As previously noted, UNESCO has played a pivotal role in the development of the right to science provision in the ICESCR. This right was significantly influenced by UNESCO's constitution and mission, which reflect the organisation's commitment to furthering the progress of science. As the leading international organisation with a mandate closely tied to these concepts, UNESCO and its activities hold relevance for understanding and safeguarding scientific freedom at the international level.

One of UNESCO's primary functions is the development and establishment of norms pertaining to its areas of expertise, achieved through various standard-setting instruments such as conventions, recommendations, and declarations. While conventions carry legal force, recommendations and declarations are considered soft law instruments, serving as guidelines rather than legally binding obligations. This distinction can sometimes prove advantageous, as soft law instruments may be adopted more swiftly and enjoy broader acceptance than formal conventions.

In the context of scientific freedom and the right to science, UNESCO's standard-setting activities are of particular importance. Two notable examples include the 2017 Recommendation on Science and Scientific Researchers and the 2021 Recommendation on Open Science. These instruments are designed to address key aspects of the conduct and furtherance of international science, including scientific and academic freedom, and are viewed by many domestic and international civil servants as providing valuable guidance for the international community.

In the following sections, the Recommendations on Science and Scientific Researchers and the Recommendations on Open Science will be introduced, with a focus on their content relating to scientific and academic freedom. Subsequently, these instruments will be briefly analysed and critiqued from the perspective of the scientific freedom framework developed earlier and summarised in the SAFIRES model. A more detailed assessment will follow in the next chapter of the present book.

2017 Recommendation on Science and Scientific Researchers

The 2017 UNESCO Recommendation on Science and Scientific Researchers is a UNESCO standard-setting instrument designed to address the evolving needs and challenges faced by the scientific community and to promote international scientific cooperation.[31] Its origins lie in a series of regional meetings in the 1960s aimed at promoting international scientific cooperation and applying science and technology to further member states' development.[32] UNESCO's experience with these meetings led it to initiate a series of expert meetings and consultations with member states on the desirability of an international instrument on the topic.[33] The eventual result was the 1974 Recommendation on the Status of Scientific Research Workers.[34]

In response to worries about the continued relevance of this 1974 Recommendation, a decision was made to update the document to address changes in the conduct, impact, and social aspects of science, including diversity and gender issues, climate change, advances in information and communications technology, and concerns over the dual use or misuse of science and technology.[35] As a result, the revised 2017 Recommendation carries over much of the content of its predecessor while including many topics of increased contemporary salience and relevance.

The 2017 Recommendation on Science and Scientific Researchers mentions the right to science as well as the connection between scientific freedom and scientific progress in its preamble,[36] and the connection between scientific progress and human interests in the main body of its text.[37] Among 'the conditions that can deliver high-quality science in a responsible manner,' it lists an 'intellectual freedom which should include protection from undue influences on their independent judgement.'[38] It should be fully taken into account, moreover, 'that creativity of scientific researchers should be promoted in national policy on the basis of utmost respect for the autonomy and freedom of research indispensable to scientific progress.'[39]

The 2017 Recommendation furthermore characterises freedom to access science and its applications 'as not only a social and ethical requirement for human development, but also as essential for realizing the full potential of scientific communities worldwide.'[40] From this follows its suggestion that states should 'put in place policies aiming to facilitate that the scientific researchers freely develop and contribute to sharing data and educational resources.'[41] The Recommendation makes clear that it also envisions positive steps. It recognises the importance of science education and appropriate employment conditions for ensuring equal participation in science.[42] States 'should provide material assistance, moral support and public recognition conducive to successful performance in research and development by scientific researchers.'[43]

Despite these areas of overlap, the 2017 Recommendation is the document that most diverges from our analysis of scientific freedom among our three international instruments. It contains numerous provisions aimed at burdening science with bureaucracy, metrics and performance appraisals. Its definition of 'science' is narrower than that advanced here. It also makes recommendations in areas of peer and ethical review, as well as with respect to intellectual property protections, that are too specific or demanding and therefore risk undermining scientific freedom. These issues are explored further in the next chapter.

2021 Recommendation on Open Science

Apart from the General Comment, the 2021 Recommendation is the document that most closely pertains to scientific freedom as a constitutive element of the right to science. Open science has various definitions and dimensions, but centrally concerns increasing access and participation in science, and thus scientific progress, by means of making easily and freely available and accessible scientific data, information, and equipment. The concept and practice of open science thus has significant affinities with the right to science, scientific freedom, and the aims of UNESCO as expressed in its constitution, among them the facilitation of scientific exchange and the wide diffusion of educational, cultural, and scientific materials.[44]

For these reasons, UNESCO has long been active in open science and adjacent areas. Among its standard-setting activities are the 1999 UNESCO/ICSU Declaration on Science and the Use of Scientific Knowledge[45] and the 2019 UNESCO Recommendation on Open Educational

Resources (OER).[46] The 2021 Recommendation is the latest of these initiatives as well as the one most closely related to scientific freedom and the right to science.

The 2021 Recommendation affirms the right to science and ties open science practices to scientific progress and human benefit in its preamble.[47] Although it primarily mentions academic, rather than scientific, freedom, it recognises that concept as a core value of open science.[48] It also recognises participation in science as a guiding principle.[49] The Recommendation recognises that support for open science requires going beyond refraining from interference to more positive steps such as 'systematic and long-term strategic investment in science technology and innovation, with emphasis on investment in technical and digital infrastructures and related services, including their long-term maintenance.'[50] More so than the other two instruments, the 2021 Recommendation emphasises the need for scientific infrastructure to support scientific freedom and open science principles, including by investing in human capital via education in and dissemination of science.[51] Despite the focus on academic, rather than scientific freedom, the Recommendation makes it clear that states should facilitate the participation of citizen scientists.[52]

Like the other two instruments, the 2021 Recommendation thus accords with the understanding of scientific freedom developed here to a significant extent. However, as is the case for the other two instruments, the 2021 Recommendation also contains areas in tension with that understanding. For example, making use of both a narrow and a wide conception of 'science,' it is unclear as to the scope of its protection. Its usage of 'academic' instead of 'scientific' freedom also diverges with our understanding of these terms. The instrument furthermore countenances wider grounds for limitation of scientific freedom than those considered acceptable under the SAFIRES model. Like the other two instruments, the 2021 Recommendation envisages public participation in scientific agenda setting. Although it admirably recommends moving away from metrics and appraisals based on bibliometrics such as citation counts and impact factors, it replaces these with open science metrics.

The 2021 Recommendation and the other two soft-law instruments are explored in greater detail in the following chapter.

Chapter 6

The Scope of 'Science' and 'Scientific Freedom' in Three Human Rights Instruments

The present chapter applies the SAFIRES model to three international instruments of great relevance to scientific freedom as well as to science and society more broadly: The 2020 CESCR General Comment, the 2017 UNESCO Recommendation on Science and Scientific Researchers, and the 2021 UNESCO Recommendation on Open Science. While the provisions in all three instruments largely cohered with our understanding of scientific freedom, there were areas of tension. We explore these areas of tension below. The structure is thematic. Each part of the chapter explores a particular theme and how this theme is treated in the three instruments. These themes are the 'purposes and aims of science,' 'performance evaluation, metrics, and accountability' and 'participation in science (decision-making)' to 'IP protection,' 'the requirement of peer review,' 'ethics review' and 'limitation criteria.'

Applying the SAFIRES model to the three UN instruments, we point to areas of concern for us—that is, to concrete thematic treatments and underlying attitudes in these instruments that are in tension with the point of view of scientific freedom as a constitutive element of the right to science advanced in this book. Our intention is to provide practical examples of the model's usefulness and its broader implications for collaboration between fields, the scope of scientific freedom protections, and the overall advancement of knowledge. We end with an extended critique of the definition of 'science' in the three instruments.

PURPOSES AND AIMS OF SCIENCE

To varying extents, each of the three instruments we focus on in this chapter envisions or defines purposes for science. These are aims and goals which, in the view of the instruments, the development of science should seek to promote. Thus, the General Comment notes that 'the development of science in the service of peace and human rights should be prioritized by States over other uses.'[1] It goes on to specify the needs of people living with poverty as an area particularly worthy of prioritisation.[2] Although not expressed in terms of prioritisation, the General Comment additionally emphasises a gender-sensitive approach as 'a crucial tool' which must be 'incorporated from the first stage, such as the choice of the subject and the design of methodologies, and must be present throughout all steps of scientific research and its applications, including during the evaluation of its impacts. Decisions concerning funding or general policies must also be gender-sensitive.'[3]

The 2021 Recommendation's preamble opens by mentioning several specific global challenges, including climate change, inequality, and humanitarian crises. It then points to the crucial role of science and technology in addressing these but does so without tying the development of science to these ends. The Recommendation does, however, propose assigning goals and priorities centered on inclusion and prioritising the needs of the least well off to the development of science. The preamble mentions several of these priority groups, including 'women, minorities, indigenous scholars, scholars from less-advantaged countries and low-resource languages.'[4] Unlike the General Comment, the 2021 Recommendation explicitly justifies the assignment of these goals by tying inclusion and participation to scientific progress.[5]

The 2017 Recommendation on Science and Scientific Researchers is more direct than the two other instruments; at several points, it refers to science as an instrument to achieve important goals. Thus, Beiter notes that:

Overall, the Recommendation reflects a rather instrumental conception of science. States are required to ensure that science

> 'tackl[es] various world problems,' 'strengthen[s] co-operation among nations,' 'promot[es] the development of individual nations,' helps 'set[ting] up a society that will be more humane, just and inclusive,' 'enhance[s] . . . the cultural and material well-being of its citizens,' 'further[s] the United Nations ideals,' and promotes 'the achievement of national goals.' 'Research and development is not [to be] carried out in isolation.'[6]

In all three cases, the assignment of externally imposed goals to scientific development, however desirable, can be seen as potentially problematic from the perspective of scientific freedom. In our understanding, as outlined in previous chapters, scientific freedom entails the right to freely pursue research without undue restrictions. This understanding is based on the belief that such freedom fosters the advancement of knowledge, and that assigning specific goals and priorities may constrain scientific freedom by directing research towards particular ends. Here, we follow the drafters of the right to science provision who thought that science must be left free as far as possible to develop according to its own dictates. By opposing the imposition even of goals such as furthering peace and democracy, as we saw in Chapter Three of Part I, these drafters made it clear that they perceived the imposition of any goals, even praiseworthy ones, to be a restriction on scientific freedom.

In terms of the SAFIRES model, the imposition of goals and priorities negatively affects the enjoyment of one or more of the elements of scientific freedom, especially freedom from interference. These priorities must therefore meet the ICESCR Article 4 ICESCR limitations criteria to be legitimate. This means that, in addition to respecting the rule of law, any limitation must be necessary for the general welfare in a democratic society and be compatible with the nature of the ICESCR rights. The latter requirement is understood here as meaning that any limitation must be appropriate and proportional, and compatible with the minimum core obligations of the right to science.

Applying these criteria to the goals and priorities imposed on science by the three instruments entails questioning whether they are strictly necessary for the general well-being, appropriate and proportionate to the aim pursued, and compatible with the core obligations of the right to science. While this is a difficult analysis requiring value judgements that may well be contested, we believe that crucial distinctions between the goals and priorities in the various instruments can be made—or, at the very least, that an attempt to make such a distinction may be instructive and hence worthwhile.

Both the General Comment and the Recommendation on Open Science involve a prioritisation of the inclusion and needs of the least well off. The Recommendation on Open Science, moreover, justifies the assignment of these goals by tying inclusion and participation to scientific progress: by promoting a more diverse and inclusive scientific community, the quality, reproducibility, and impact of science are expected to improve. To the extent that this justification is true, by fostering a more inclusive and diverse scientific community, open science practices can simultaneously advance scientific progress and address pressing societal needs without compromising scientific freedom. In this light, the UNESCO Recommendation on Open Science reconciles the goals of inclusion and the needs of the least well off with the principles of scientific freedom, providing a balanced approach to the development of science.

While the emphasis on participation thus seems unproblematic from the perspective of the limitation criteria, the situation is more complicated when it comes to the development of science in areas expected to materially improve the situation of the least well off, as opposed to including them in the scientific process. Here, it may be a good idea to distinguish between the creation of scientific knowledge itself and the practical use to which this knowledge is subsequently put—or, in the language of ICESCR Article 15(1)(b), between the 'enjoy[ment of] the benefits of scientific progress and its applications.'[7] There are strong reasons to impose priorities of inclusion on the 'application' part of Article 15(1)(b) – that is, on the distribution of the benefits and applications that result from scientific progress, including technologies and knowledge. Later on in this chapter, we shall discuss the issue of dual-use science and technology, for example—an area where it is crucial for the general public to be included in decisions concerning scientific 'products.'

In terms of the development or production of scientific knowledge itself, however, things are a bit more complex. As highlighted by the epistemic argument surveyed in Chapter One of Part I as well as by the drafters of the right to science, it is difficult to predict in advance what shape the scholarly development of science will take. For this reason, when it comes to 'scientific progress' it may be better to allow serious scientists to experiment and try out various methods and approaches based on their individual expertise and experience, rather than require them to follow a predetermined path. In this context, it is worth stressing, as we have in previous chapters, that from a human rights perspective, scientific freedom does come with responsibility.

As argued in Chapter One, Part I, there may be strong reasons to impose goals on urgent research in the short term and pragmatic and democratic reasons to foster national priorities in mission-driven research in the medium term. In these cases, the epistemic argument suggests that governments may be sacrificing some overall scientific progress (by diminishing scientific freedom) in return for advancement on a particular area of priority, such as vaccine development. Accepting such a trade-off may make sense in specific contexts, i.e., when a response is needed to an emergency[8] and/ or to addressing problems that are important but too obscure to catch the interest of scientists (e.g., a rare disease). Somewhat problematically, however, the 2017 Recommendation contains a set of aims and goals related to non-urgent national goals, such as economic development. According to the Recommendation, 'Member States should treat public funding of research and development as a form of public investment [. . . and] ensure that the justification for, and indeed the indispensability of such investment is held constantly before public opinion.'[9] Science should also "enhance . . . the cultural and material well-being of its citizens"[10] and promote 'the achievement of national goals.'[11]

While these passages do not explicitly say so, when combined with requirements to introduce performance evaluations and metrics (see section below), they come quite close to asserting that economic considerations such as profit should guide the direction of research. There are at least three reasons to be careful of introducing profit motives into research, especially in academic settings. First, it has long been understood that the primary functions of academic science are teaching and the pursuit of knowledge.[12] There is nothing inherently wrong with profit-motivated research, but this is a separate pursuit that should not compete with knowledge-motivated research, and which should take place in companies and research institutions not primarily dedicated to the disinterested advancement of knowledge. Second, and relatedly, the introduction of a profit motive can conflict with and therefore undermine the disinterested pursuit of knowledge, as is well known from research on biases introduced by conflicts of interest in biomedical research.[13] Finally, there is reason to believe that explicitly pursuing economic goals might be counterproductive, in the sense that the most economically impactful advances have often come from undirected, or minimally directed, blue-skies research rather than from commercial research.[14]

While the tension between freedom of science and priority setting is not easy to resolve, three conclusions can be drawn from the above discussion. First, the development of science should be as free as possible of external goals. Second, where their imposition is necessary because of emergencies such as pandemics or wars, these goals should be restricted to the highest level of generality and should not dictate specific methods or approaches of investigation. Finally, even where prioritisation is required to meet pressing needs, funding should still be available for entirely goal- and priority-free research. No matter how urgent a priority or emergency is, states should ensure that a minimum level of independent funding is available for fundamental, goal- and priority-free research.

PERFORMANCE EVALUATION, METRICS AND ACCOUNTABILITY

Another set of restrictions on academic and scientific freedom arises from systems of performance evaluation, metrics and monitoring. These restrictions are introduced in efforts to increase the public accountability of researchers, presumably with the intended effect of thereby increasing their efficiency and making sure that they do not waste funding. While these themes are present in all three instruments, they are significantly more pronounced in the 2017 and 2021 Recommendations than in the General Comment.

The 2017 Recommendation states that each 'Member State should institute procedures adapted to its needs for ensuring that, in the performance of research and development, scientific researchers respect public accountability

[. . .].'[15] According to the General Comment, '[a]s far as possible, scientific or technological policies should be [. . .] implemented with accompanying transparency and accountability mechanisms.'[16] And the 2021 Recommendation on Open Science lists accountability as a guiding principle for open science and mentions public accountability as a basis for its good governance.[17]

To bring about this accountability, the instruments recommend the introduction of various performance evaluation mechanisms and metrics. Most stringently, the 2017 Recommendation suggests that each Member State should not only 'establish appropriate [. . .] appraisal systems' based on 'publications, patents, management, teaching, outreach, supervision, collaboration, ethics compliance, and science communications,'[18] but also 'combine appropriate metrics with independent expert assessment (peer review) of the individual's outputs, as to all aspects of the work.'[19] The wording of the 2021 Recommendation is somewhat less emphatic. The Recommendation does state in broad terms, though, that '[a]ssessment of scientific contribution and career progression rewarding good open science practices is needed for operationalization of open science.'[20] It furthermore recommends that states should promote 'the development and implementation of evaluation and assessment systems' which 'take into account evidence of research impact and knowledge exchange.'[21] More generally, the Recommendation also wishes to ensure that 'the practice of open science [. . .] is taken into account as a scientific and academic recruitment and promotion criterion.'[22]

As with all recipients of public funding, researchers are under obligations not to misappropriate, waste or misuse the money invested in them by the public, and to make the best possible use of it. At least partly for this reason, in the academic context, numerous professional norms and codes govern research conduct. Prime among these are strictures against plagiarism and fudging or inventing data and analyses. These are mechanisms meant to weed out cheaters and those whose work consistently falls below the professional or community standards of the scientific community. Such mechanisms are efforts at self-regulation, meant to facilitate the maintenance of professional standards.

The systems of performance evaluation, metrics and accountability mentioned above, by contrast, involve external oversight of scientific conduct. This is not in principle or in itself problematic. But such oversight does become a problem when externally imposed evaluation procedures introduce biases and obstacles into research that end up counter-productively decreasing efficiency and progress. This is not a problem only for externally imposed procedures, but the lack of experience with and understanding of science and research processes on the part of people outside these processes may well intensify it.

Among the most pernicious effects of performance evaluation and metrics is the introduction of an additional motivation to the conduct of researchers: profit or performance incentives. These may distract from, conflict with, and undermine the primary function of the pursuit of knowledge. A key example is the phenomenon known as 'publish or perish.' Since scientists often work on highly specialised topics, those tasked with evaluating their performance rarely have the specific domain expertise or time required to accurately understand and evaluate their performance. Instead, proxies for scientific achievement are used. Prime among these are the number of publications and/or patents, the ranking of the journal in which these were published, and the number of times a scientist's articles or patents have been cited.[23]

As these metrics are widely used to assess research quality and impact,[24] numerous studies have investigated their effects on scientific research. Because studies that find statistically positive results are more likely to get published, pressure to publish introduces biases toward the publication of positive results. This effect, known as publication bias, is more pronounced in more competitive settings.[25] The use of publication metrics in performance review also leads to a focus on quantity over quality, resulting in a flood of low-quality and irreproducible articles in important fields such as medicine.[26] As a necessary corollary, the need to publish early in order to demonstrate progress in annual performance reviews leads to a focus on short-term projects over the tackling of more difficult but potentially more important longer-term projects. This same pressure can lead to researchers cutting corners, engaging in questionable research practices or even in outright misconduct and fraud.[27] More generally, the introduction of a system of incentives and rewards based on proxies leads naturally to behavior aimed at gaining rewards by gaming that system.[28]

Fundamentally, the conflicting incentives and pressures introduced by the reliance on metrics and performance evaluation decrease the time and motivation available for science, focusing efforts on the meeting of indicators and proxies rather than the underlying scientific progress that these proxies are meant to measure. Thus, a recent study observing a decrease in the rate of disruptive scientific research over the past several decades recommends that, to counter this trend, universities should 'forgo the focus on quantity, and more strongly reward research quality.' This includes making time available for 'keep[ing] up with the rapidly expanding knowledge frontier' and 'more fully subsidiz[ing] year-long sabbaticals.' As for funders, they should 'invest in the riskier and longer-term individual awards that support careers and not simply specific projects.'[29]

Overall, while well-intentioned, requirements for metrics and performance evaluation are likely to do more harm than good from the perspective of scientific freedom and progress. Acting as a break on scientific progress without introducing benefits that might outweigh their deleterious impacts, such requirements do not meet the limitations criteria and consequently are incompatible with the SAFIRES model and the right to science on which it is based. They therefore have no place in instruments like the 2017 Recommendation and should be removed in future revisions. The relatively smaller place and importance accorded to accountability and performance evaluations in the General Comment and the 2021 Recommendation may therefore be seen as steps in the right direction—towards scientific progress through freedom, and away from unintended consequences and perverse incentives.

An open question remains, though: are more specific metrics and evaluation criteria, focused more narrowly on open science practices such as open access publication, pre-registration of protocols, and the making available of data and detailed methods in supplementary information, subject to this same conclusion, or do they constitute an exception to it? While the introduction of requirements and incentives for open science practices may introduce the same kind of perverse incentives mentioned above, they may also further scientific progress by bringing more scientific information into the publicly accessible domain. If restricted to making research more openly accessible, such requirements might facilitate rather than impede scientific progress. However, their introduction should be carefully considered and their effects monitored for unintended consequences.

PARTICIPATION IN SCIENCE (DECISION-MAKING)

As noted, a key element of the right to science, and of scientific freedom as understood here, is participation in science. The right to participate in scientific progress implies that there should be available sufficient scientific education, training and infrastructure that those who are interested can train in and carry out scientific research. It also implies that there should be no discrimination in or unnecessary barriers to accessing this education, training and infrastructure and thus to participating in scientific progress. The 2020 General Comment and 2021 Recommendation, however, argue that the right to participate in scientific progress extends beyond these implications. According to these instruments, the right to participate in science includes the right to participate in scientific decision-making and agenda- and priority-setting.

The 2020 General Comment contains many statements to this effect. According to the General Comment, 'the right of everyone to take part in cultural life includes the right of every person to take part in scientific progress and in decisions concerning its direction.'[30] States should 'promote public trust and support [. . .] through a vigorous and informed democratic debate on the production and use of scientific knowledge, and a dialogue between the scientific community and society.'[31] Paragraphs 55 and 56 state the issue most clearly and explicitly note the tension with scientific freedom:

> With due respect to scientific freedom, some decisions concerning the orientation of scientific research or the adoption of certain technical advancements should be subjected to public scrutiny and citizen participation. As far as possible, scientific or technological policies should be established through participatory and transparent processes and should be implemented with accompanying transparency and accountability mechanisms. [. . .] Participation also includes the right to information and participation in controlling the risks involved in particular scientific processes and its applications.[32]

The 2021 Recommendation on Open Science envisions 'an enhanced dialogue between scientists, policymakers and practitioners, entrepreneurs and community members, giving all stakeholders a voice in developing research that is compatible with their concerns, needs and aspirations.'[33] It also encourages Member States to develop 'participatory strategies for identifying the needs of marginalized communities and highlighting socially relevant issues to be incorporated into the science, technology and innovation (STI) research agendas.'[34]

To what extent is popular participation in science decision-making and priority-setting compatible with scientific freedom? According to the SAFIRES model, this is when such participation either maintains or increases scientific freedom (freedom from interference, freedom to access and freedom to participate in science) or else is compatible with the Article 4 limitation criteria. Evaluating the extent to which this is the case is not straightforward, as popular participation simultaneously enhances one aspect of scientific freedom (freedom to participate) and restricts another (freedom from interference).

One way to resolve the tension is to analyze it as analogous to the tension discussed above between freedom from interference and the imposition of goals and priorities on science. Conceived in this way, popular participation in scientific policy- and decision-making either becomes a goal or priority of its own or else serves as a mechanism through which goals and priorities are identified. This serves as a limit on scientific freedom, essentially trading a reduction in the rate of scientific progress generally for faster advancement on particular topics. However, as argued above, this trade-off might be acceptable, provided the goals and priorities identified through popular participation are important, urgent, and at a sufficient degree of abstraction and generality such that they do not dictate specific methods, angles, or approaches to be used or taken in pursuing those goals. Importantly, even if such participation is sought, there should nevertheless be sufficient resources available for completely blue-skies, interest-based research entirely free of imposed direction.

IP PROTECTION

By its very nature, intellectual property (IP) protection, such as patents and copyright, removes or restricts access to scientific information, data, or equipment. It is therefore fundamentally in tension with the freedom of access element of scientific freedom under the right to science. The tension between access and IP protection is a key aspect of the literature on the human right to science, appearing in the great majority of published studies on the topic.[35] It also occupies a prominent role in all three of the international instruments examined here.

The 2017 Recommendation makes numerous references to access and IP protections, among them the importance of 'fully respect[ing] the intellectual property rights of individual researchers'[36] and 'ensur[ing] that the scientific and technological results of scientific researchers enjoy appropriate legal protection of their intellectual property, and in particular the protection afforded by patent and copyright law.'[37] These mentions sit side-by-side with exhortations to increase the openness and accessibility of science, forcing the 2017 Recommendation to make relatively bland statements of the need for 'balancing between protection of intellectual property rights and the open access and sharing of knowledge.'[38]

The General Comment clearly notes the tension between IP rights and access to science and its applications: 'On one hand, intellectual property enhances the development of science and technology through economic incentives for innovation, such as patents for inventors, which stimulate the involvement of private actors in scientific research. On the other hand, intellectual property can negatively affect the advancement of science and access to its benefits.'[39] Of the three instruments, the General Comment is most forceful in pushing back against excessive IP protections:

A balance must be reached between intellectual property and the open access and sharing of scientific knowledge and its applications, especially those linked to the realization of other economic, social and cultural rights, such as the rights to health, education and food. The Committee reiterates that ultimately, intellectual property is a social product and has a social function and consequently, States parties have a duty to prevent unreasonably high costs for access to essential medicines, plant seeds or other means of food production, or for schoolbooks and learning materials, from undermining the rights of large segments of the population to health, food and education.[40]

This respect for IP rights is noticeable even in the 2021 Recommendation on Open Science, which is dedicated to the very openness and accessibility challenged by IP protection. The preamble of the 2021 Recommendation 'recogniz[es] the importance of the existing international legal frameworks, in particular on intellectual property rights including the rights of scientists to their scientific productions.'[41] Its main text even envisions 'the protection of intellectual property rights' as a legitimate reason for limitation of open access to scientific knowledge.[42]

As argued above, under the conception of scientific freedom developed here, limitations on scientific freedom—including limitations on access to the necessary inputs to the scientific process—must be justified according to specific criteria. Since the underlying reasoning for IP protection is, in part, to motivate scientific discoveries, they are surely compatible with the nature of the rights contained in the ICESCR, which includes the right to science itself. However, nearly the entire extant right to science literature questions whether the current extent of protection is strictly necessary for the general well-being, arguing mostly that such protection is too high.[43] Therefore, the Recommendation should not recommend, from the point of view of scientific freedom, that states accept the status quo.

To their credit, the instruments do not state that current IP regimes are optimal. However, apart from the General Comment, they also do not do much in the way of fighting for greater access beyond the numerous references to the importance of doing just that as illustrated by the provisions quoted above. Beiter, commenting on the 2017 Recommendation, points out that the main problem here is one of vagueness:

> The Recommendation does call for 'balancing between protection of intellectual property rights and the open access and sharing of knowledge.' It should, however, also have included specific suggestions for statutory limitations and exceptions to copyright protection that states could implement to safeguard access to scientific knowledge in educational or scientific institutions, libraries and archives. Insofar as the intellectual property rights of scientists themselves are concerned, the Recommendation does little to add to or strengthen these.[44]

It would be better if both Recommendations could give concrete advice to states in this regard. Although the frequent references to open access, which greatly outnumber the references to IP protection, are a clear step in the right direction, the protection of scientific freedom requires more than this. If UNESCO and its instruments are not more forceful in this area, it is hard to see which organisation or groups are going to have the incentives, authority and ability to pick up the slack and provide the necessary pushback to the financial interests behind IP protections.

THE REQUIREMENT OF PEER REVIEW

Peer review—in which candidate papers for publication are vetted by journal editors and colleagues ('peers') – is a long and hallowed tradition in scientific research. Though there are instances of review processes dating back to antiquity, the first modern processes were introduced by the Royal Society of Edinburgh in 1731.[45] This procedure involved the review of submitted manuscripts by the editor and a fixed board of experts; the current procedure, in which copies of a submission are sent to external scientists with domain expertise, was not widely adopted until the mid-twentieth century.[46]

Peer review serves two core functions: assisting editorial staff to make a decision on publication, and vetting and improving the manuscript.[47] In addition, peer review is often seen as the distinguishing feature serving to legitimise or verify a manuscript as a bona fide piece of scholarship.[48] As a result, peer review is routinely viewed as a 'gold standard' and has become one of the strongest social constructs in academic self-regulation.[49] Its legitimising function has only become more important given the rise of substandard and predatory journals.[50]

It is thus understandable that requirements for peer review feature in the international instruments dedicated to science. According to the 2017 Recommendation, 'Member States should establish as a norm for any scientific publishing, including publishing in open access journals, that peer review based on established quality standards for science is essential.'[51] Peer review even features in the 2017 Recommendation's definition of 'science': 'the word 'science' signifies the enterprise whereby humankind [. . .] by means of the objective study of observed phenomena and its validation through sharing of findings and data *and through peer review*, to discover and master the chain of causalities, relations or interactions [. . .].'[52]

However, the process of peer review has been subject to criticism since its introduction.[53] Despite its critical role, peer review faces several acknowledged challenges, including inadequate training and support for researchers, lengthy review processes and the loss of valuable contextual information due to unpublished review reports. Moreover, finding suitable operational processes for different research communities, addressing the lack of rigorous evidence on peer review effectiveness, and clarifying the relationship between review quality and journal quality are ongoing concerns.[54] A recent review found seven categories of potential bias introduced by peer-review processes, ranging from inability or unwillingness to properly assess the quality of studies through conservatism and confirmation bias to conflicts of interest.[55]

To address these known issues, many variations of peer review have been proposed. These introduce different degrees of openness into the system, from quadruple-blind review (in which the identities of the editorial board, handling editor, author, and peer reviewers are all kept secret) through triple-, double-, single-blind and fully open peer review. The motivation behind blinded reviews is to reduce biases introduced by reviewers knowing the identity of an author, while the motivation behind open reviews is to decrease delays and biases introduced by the knowledge of a peer reviewer that their work will not be associated with their name. To complicate matters further, there are also editor-only and post-publication reviews; and whole fields of science—most prominently mathematics, computer science and physics, though biology and medicine are also trending in this direction—have begun publishing manuscripts or preprints directly, without peer review, to online 'preprint' archives.

In response, the 2021 Recommendation on Open Science recommends 'extending the principles of openness in all stages of the scientific process to improve quality and reproducibility, including the encouragement of community-driven collaboration and other innovative models, for example preprints, clearly distinguished from final peer-reviewed publications,'[56] and '[p]romoting, as appropriate, open peer-review evaluation practices including possible disclosure of the identity of the reviewers, publicly available reviews and the possibility for a broader community to provide comments and participate in the assessment process.'[57]

In addition to possible biases, peer review necessarily introduces costs and delays. Given the costs, lack of evidence of best practices to discriminate between the many variations, known biases and delays, and different approaches to review between scientific fields, the recommendations and requirements in the 2021 and 2017 Recommendations are too specific. Too little is known about the relative merits of open and closed peer review to mandate the use of either in science. There is also too much heterogeneity between fields of science regarding factors such as the cost and impact of delays and the degree and severity of competitive pressures leading to biases to impose the use of any one system of peer review on all fields of science. While the motivation behind these recommendations is benevolent, seeking to address biases and distortions such as those arising from predatory publishing, there are safer ways to do so, such as moving away from the publish and perish mentality mentioned above.

For these reasons, the approach of the General Comment on this matter is to be preferred. While the Comment contains no mention of peer review as a specific practice, it speaks of 'regulation and certification, as necessary,' and 'the most advanced, up-to-date and generally accepted and verifiable science available at the time, according to the standards generally accepted by the scientific community' as determinants of quality.[58] This language is broad enough to cover many variations of peer review as well as other methods of safeguarding and improving the quality of scientific publication and dissemination.

ETHICS REVIEW

Similar points to those above on peer review may be made about the requirement for ethical review of human subjects research by ethical or institutional review boards (IRBs). This requirement is the result of public condemnation of and reaction to specific instances of unethical research, particularly dangerous and deceptive medical experiments carried out without the informed consent of experimental subjects. Instances of such reaction date back at least to the early nineteenth century; in 1891, the Prussian government was the first to institute legal consent requirements for participation in medical research.[59] However, the requirement became widespread following the Nuremberg trials of medical atrocities during the Second World War and was codified in legal and professional, in addition to bioethical, codes following the release of high-profile books and reports on unethical medical experiments in the UK and US by Maurice Papworth and Henry Beecher, respectively, in the late 1950s and 1960s.[60] These codes were revised and expanded following the publication of the *Belmont Report* in 1978 in response to a particularly severe breach of medical ethics in the Tuskegee Syphilis Study.[61]

The purpose of ethics boards is primarily to protect human research subjects from harm. Much of biomedicine, for example, involves speculative interventions based on theory and incomplete data into the complex biological reality of the human body. Since scientists working in these fields are attempting to push beyond existing knowledge, there is often significant uncertainty involved in medical trials; if this were not the case, there would be no point in conducting such trials. However, the trials can be very dangerous to participants, and therefore require strong justification. Ethical review boards are tasked with ensuring that the benefits to be expected from these trials outweigh the likely or possible negative consequences for participants, and with ensuring that these participants have been fully informed of the relevant risks and benefits and have given their fully voluntary consent to be a part of the study.

It is thus understandable that requirements for ethics review are found in international instruments related to science. IRB review features prominently in the 2017 Recommendation, which urges States to:

[establish] suitable means to address the ethics of science and of the use of scientific knowledge and its applications, specifically through establishing, promoting and supporting independent, multidisciplinary and pluralist ethics committees in order to assess the relevant ethical, legal, scientific and social issues related to research projects involving human beings, to provide ethical advice on ethical questions in research and development, to assess scientific and technological developments and to foster debate, education and public awareness and engagement of ethics related to research and development.[62]

The General Comment refers to ethics and ethics review in several places but uses slightly less mandatory language. The Comment recommends 'measures to ensure that ethics and human rights are respected in scientific research, including the establishment of committees on ethics when necessary.'[63] It also suggests a national plan of action including 'measures to ensure ethics in science, such as the establishment or promotion of independent, multidisciplinary and pluralist ethics committees to assess the relevant ethical, legal, scientific and social issues related to research projects.'[64] For reasons discussed shortly, the General Comment's insistence on consent for human subject research without exception is also noteworthy.[65]

There is little doubt that mechanisms should be in place to ensure that human subjects are not used against their will in dangerous experiments for the gain of others. However, the imposition of ethical review for all human subjects is an imperfect mechanism for several reasons. Chief among these is that it is doubtful whether these committees have been given an appropriate remit and whether they are fit for this purpose.

Ethical review is currently required for all human subjects research in several countries. Since review can take anywhere from months to years, this imposition necessarily involves opportunity and temporal costs—not to mention the oft-high financial cost involved in obtaining consent—which translate into delays to research. These in turn translate into lives not saved and illnesses not cured.[66] Whereas few would doubt the necessity of ethical oversight for highly interventional research on procedures like surgery or chemo- or radiotherapy, the costs do not necessarily outweigh the benefits for whole classes of research which cannot be characterised in this way. For example, much of social sciences research involves either routine surveys or interviews with human subjects, and epidemiological research on pre-existing data is also considered to be human subjects research in practice if not always in theory. Thus, the criticism can be made that it is crass to subject a student project in which family members are asked about their interior design preferences to the same legal requirements as an experimental neurosurgery protocol.[67] It is for these reasons that the blanket requirement of consent in the General Comment goes too far and risks undermining beneficial research by imposing costly consent requirements on minimally risky or risk-free research.

Secondly, doubts may be raised as to whether ethics review boards are successful at their mandate. Since each research institution (under US law) or region is required to have an ethics board, comparisons can be made between their judgements. Studies that have compared identical or highly similar protocols have found large variability in ethics review.[68] This observation continues to be made in the context of multicentre trials, in which the local IRBs of each participating institution reviews study protocol, often with widely varying results.[69] Although the idea of a multidisciplinary ethical review board is good, in practice, review boards are often made up largely of doctors and scientists, and very rarely does any member of the team have any formal education in clinical or research ethics or even in ethics at all. Often, they are junior members 'volunteering' to do the work, which is typically unpaid and not taken into account for purposes of career progression.[70] Finally, the observation has often been made that, at least in cultures with a heavy tradition of medical litigation, the de facto primary aim of IRBs appears to be shielding their institutions from liability.[71] Where this is the case, it results in a highly skewed appraisal of protocols in general, and a disproportionate focus on consent forms in particular.

Like peer review, ethical review is an imperfect solution to a real problem. Often it is necessary, but sometimes it is not. When the latter is the case, it leads to foreseeable costs and delays which are not necessary for the general well-being and thus unjustifiably restrain scientific freedom. Furthermore, it is doubtful whether the current system is optimised to fulfil this mandate. Thus, ethical review should not be required or even recommended for all human subjects research as it appears to be in the 2017 Recommendation and the General Comment.

LIMITATION CRITERIA

While the General Comment acknowledges the Article 4 ICESCR limitation criteria, the 2021 Recommendation on Open Science goes beyond these.[72] It states:

> Access to scientific knowledge should be as open as possible. Access restrictions need to be proportionate and justified. They are only justifiable on the basis of the protection of *human rights, national security, confidentiality, the right to privacy and respect for human subjects of study, legal process and public order, the protection of intellectual property rights, personal information, sacred and secret indigenous knowledge, and rare, threatened or endangered species.*[73]

As mentioned above, the only legitimate purpose for limitations on rights in the ICESCR is the general welfare. Other human rights treaties, such as the International Covenant on Civil and Political Rights (ICCPR), envision broader grounds for limitations. Article 19(3) ICCPR, for example, recognises 'respect of the rights or reputations of others, for the protection of national security, public order, or public health or morals' as potentially legitimate grounds for limitations on freedom of expression and related rights in Article 19(2) ICCPR.[74]

The possibility of recognising broader grounds for legitimate limitations was discussed and specifically rejected during the drafting of the ICESCR.[75] Thus, the protection of open access to scientific knowledge under the right to science should be subject to the general limitation criteria in Article 4 ICESCR, but not to the wider grounds mentioned in the 2021 Recommendation, unless these can be shown, in turn, to be necessary for the general welfare.[76]

CONCLUDING REMARKS: THE SCOPE OF 'SCIENCE'

As noted in previous chapters, there are compelling reasons to interpret 'science' broadly in the context of international human rights law. The narrower usage, specific to the English language, typically refers to the natural sciences and some social sciences. However, the terms in other authentic language versions of the UDHR and ICESCR are more inclusive, reflecting the way in which the drafters of these foundational instruments saw 'science' as a part of culture, broadly speaking. Scientists seek knowledge and understanding, and it is their intellectual approach to, rather than the subject matter of, their research that determines if it is scientific. What mattered to the drafters is whether a sincere, bona fide attempt is made at advancing knowledge, rather than merely proposing a particular opinion.

The definition of 'science' used in all three of the international instruments assessed in this chapter echoes that of the English language usage rather than the broader view expressed in non-English language versions of the UDHR and ICESCR. Both the 2020 General Comment and the 2021 Recommendation on Open Science build on the definition of 'science' contained in the 2017 Recommendation on Science and Scientific Researchers:

> The word 'science' signifies the enterprise whereby humankind, acting individually or in small or large groups, makes an organized attempt, by means of the objective study of observed phenomena and its validation through sharing of findings and data and through peer review, to discover and master the chain of

causalities, relations or interactions; brings together in a coordinated form sub-systems of knowledge by means of systematic reflection and conceptualization; and thereby furnishes itself with the opportunity of using, to its own advantage, understanding of the processes and phenomena occurring in nature and society.[77]

Beyond the instruments examined here, this definition has been adopted by numerous scholars[78] and by other UN special procedures.[79] It thus reflects an emerging consensus definition of 'science.' As can be seen by the reference to 'nature and society' at the end of the definition, the definition's subject matter is meant to be the natural and social sciences.

In our view, this development is problematic. Under the definition of science in the 2017 UNESCO Recommendation, the understanding of scientific freedom and the SAFIRES model developed in this book would apply to anyone carrying out serious, systematic intellectual tasks in the natural and social sciences, but not to those carrying out equally rigorous pursuits in the humanities.[80] The definitions of 'science' and 'scientific freedom' in the 2017 Recommendation narrow the scope of protections and obligations based on the right to science, including scientific freedom. Intriguingly, this narrower usage, which includes the natural and parts of the social sciences only, may well be the result of a historical accident and a deference to the authority of UNESCO, rather than to the merits of the definition itself.

The 2017 UNESCO Recommendation on Science and Scientific Researchers' definition of 'science' originates from a working document, prepared by the UNESCO Secretariat for the first Conference of Ministers of the European Member States responsible for Science Policy (MINESPOL I), which took place in 1970.[81] This conference was part of a series of regional ministerial conferences that was initiated in the 1960s to engage with science and to further international scientific cooperation. In its synopsis, the 1970 working document suggests definitions of, e.g., science, technology and scientists, 'which will provide a sound and reliable basis for discussion, thus avoiding time being wasted on sterile semantics.'[82] The definition adopted for 'science' specifically focuses on the natural sciences:

SCIENCE may be defined as mankind's organized attempt, through the objective study of empirical phenomena, to discover how things work as causal systems. By means of systematic thought, expressed essentially in the symbols of mathematics, it brings together the resultant bodies of knowledge in an effort to reconstruct the world a posteriori by the process of conceptualization. Its purpose is not to invent but to comprehend. The sciences thus constitute an interlocking complex of attested fact and speculative theory, with the essential proviso that theories must be capable of being tested experimentally.[83]

The obvious similarity to the definition of 'science' in the 2017 Recommendation stems from the fact that when drafting its predecessor, the 1974 Recommendation on the Status of Scientific Research Workers, the UNESCO Secretariat believed creating a new definition was unnecessary. They argued that the MINESPOL I definition had proven practical by serving as a foundation for ministerial exchanges and had received member states' endorsement.[84] Consequently, the Secretariat built upon MINESPOL I's work and used the existing definition in their Recommendation, though they made clear that the definition and thus the scope of the 1974 Recommendation was meant to extend also to the social sciences. The 2017 Recommendation made only slight changes to the definition contained in the 1974 version.

This definition has since been adopted by a number of scholars and by UN special procedures and instruments beyond the ones assessed in this chapter. The very fact that this emerging consensus definition of 'science' seems to stem from initially practical and pragmatic considerations centering around the natural sciences only, and that these have then been perpetuated by deference to the authority of UNESCO and subsequent path dependence does not automatically invalidate it. However, considering the arguments speaking in favor of adopting a broader definition of 'science,' the current definition's adequacy in addressing the diverse nature of scientific disciplines and their evolving roles in society is questionable.

Achieving a principled and clear consensus on the definition of 'science' is even more important given the inconsistent applications and uses of the term, even within instruments. The 2021 UNESCO Recommendation on Open Science adopts the 2017 definition in its paragraph 4.[85] Yet just two paragraphs later, it defines 'open science' as covering 'all *scientific disciplines and aspects of scholarly practices*, including basic and applied sciences, natural and social sciences *and the humanities.*'[86] Likewise, the 2021 UNESCO Recommendation on the Ethics of AI mentions, in its Article 3b, '[s]cience in the broadest sense and including all academic fields from the natural sciences and medical sciences to the social sciences *and humanities.*' A bit further on, Article 110 mentions 'disciplines *other than science*, technology, engineering and mathematics (STEM), *such as* cultural studies, education, ethics, international relations, law, linguistics, philosophy, political science, sociology and psychology.'[87]

The historical context and the somewhat accidental way in which the definition of 'science' was arrived at in UNESCO, as well as the inconsistency with which it has since been applied in UNESCO instruments indicate that a reappraisal of this definition is warranted. If, or when, this happens it is important to look at scientific freedom as a constitutive part of the right to science from the perspective of the scientific community. Much contemporary scholarship on the issues involved here explores citizen science, the anticipation of dual-use science and technology along with the need for the external regulation of science, as well as the importance of everybody's participation in the creation of science and in science policymaking. The focus is, in other words, typically on the access to and participation in the right to science by the public. The role played by the scientific community is sometimes toned down—just as it is downplayed that without scientific freedom, no scientific knowledge can be produced from which societies may benefit in attempting to deal with global challenges such as climate change, pandemics like COVID-19, and the ever-widening gap between rich and poor.

When we first discussed among ourselves the possibility of writing a book on scientific freedom, we thought that it might be interesting to add to our own discussions the perspectives of people working in the field with scientific freedom and right to science-related issues. We therefore approached four colleagues who graciously agreed to contribute to our book. Part III contains their contributions. In Chapter One, Mikel Mancisidor shares with us his thoughts 'On Drafting the General Comment on Science (2013–2020): A Personal Account.' Cesare Romano tells the story of 'Litigating the Right to Science before the CESCR: The View from the Trenches' in Chapter Two. And finally, in Chapter Three, Malene Nielsen and Carsten Staur write about 'Defending Science, Knowledge, and Facts: The UN and Scientific Freedom of Expression.'

PART III

Working with Scientific Freedom and the Right to Science in Practice

Chapter 7

On Drafting the General Comment on Science (2013–2020): A Personal Account

Mikel Mancisidor

In this article I will first describe the Committee on Economic, Social and Cultural Rights' (CESCR) work on drafting the General Comment No. 25 on *Science and Economic, Social and Cultural Rights* (2013–2020).[1] Then, I will show the impact that this instrument is having around the world. Being one of the rapporteurs of this General Comment, and the only one who participated actively from the beginning to the end of the process, it will be difficult for me—and I have no intention to resist this particular temptation—to avoid some personal or subjective approaches.

I will start with a short historical introduction which has two parts: a very brief account of the history of the so-called right to science, and a more recent account of the approval, drafting, steps and adoption of General Comment No. 25. Then, I will discuss the legal status in international law of General Comments in general, and next, I will review very briefly a few important substantive elements of the General Comment, including scientific freedom. Finally, I will present, in a very provisional manner, some outcomes or consequences of the adoption of this General Comment, both at the international and the domestic levels. This will also serve as a conclusion concerning the experience of the drafting (2013–2020), adoption (2020) and advocacy (2020–2022) of General Comment No. 25.

A VERY SHORT HISTORY

1.1 Antecedents

The history of the human right to science, an issue ignored and forgotten for many years, has been told several times during the last fifteen years,[2] so I will not go into detail here, but just list, in a very cursory manner, some key moments.

The complex set of freedoms and entitlements[3] related to science entered both into, first, the American Declaration on the Rights and Duties of Man (1948)[4] and the Universal Declaration of Human Rights[5] and then into the International Covenant on Economic, Social and Cultural Rights (1966).[6] From that moment on, some important milestones can be identified.

A joint initiative between UNESCO, the Amsterdam Center for International Law, and the Irish Centre for Human Rights signaled the beginning of the return of the right to science to the international arena. These three institutions organised three experts' meetings in Amsterdam (2007), Galway (2008) and Venice (2009), respectively, which provided an essential basis for further advancements.[7]

Farida Shaheed, Independent Expert in the field of cultural rights, presented a thematic report in 2012 on this right,[8] after a participatory process with numerous states, scholars, and scientific and social organisations from around the world. In her report, she incorporated twenty-four points focused on the normative content of the right. This report constitutes a basic corpus essential for all future efforts:

> the normative content of the right to benefit from scientific progress and its applications includes (a) access to the benefits of science by everyone, without discrimination; (b) opportunities for all to contribute to the scientific enterprise and freedom indispensable for scientific research; (c) participation of individuals and communities in decision-making; and (d) an enabling environment fostering the conservation, development and diffusion of science and technology.[9]

Shaheed ended her report by recommending that 'the Committee on ESCR review article 15 of the Covenant in a comprehensive manner, and envisage adopting a new general comment encompassing all rights recognized therein.'[10]

As a consequence of this report, and following the mandate of the Council of Human Rights,[11] in October 2013, the Office of the United Nations High Commissioner for Human Rights organised a two-day seminar on the Right to Enjoy the Benefits of Scientific Progress (henceforth REBSP) in Geneva.[12] Several voices joined at this event to provide greater clarity on the content and scope of the new General Comment of the UN Committee on Economic, Social and Cultural Rights (henceforth CESCR).

1.2 The General Comment

Finally, in November 2013, the CESCR approved at its 51st session that work could commence on a future General Comment on the REBSP.[13] This General Comment would determine the normative content of this right and lay down guidelines for States Parties on how to comply with it and on how to facilitate their information obligations vis-à-vis the Committee.

Jaime Marchán, a very experienced member from Ecuador and former president of the Committee, and the author of the present text were appointed as rapporteurs. After Marchán left the Committee (in 2014) and until 2018 when the Committee appointed a second rapporteur, Rodrigo Uprimny from Columbia, I was left with the responsibility of advancing the GC on my own. Uprimny gave the process a new and definitive impetus thanks to both his legal and intellectual competence and his personal authority within the Committee.

From 2014 until the very last day of the sixty-seventh session (17 February–6 March 2020), Uprimny and I presented various drafts. The new GC was finally adopted on the 6th of March 2020.[14] It is worth noting that this adoption took place just five days before the WHO declared COVID-19 a pandemic.[15]

On the 9 October 2018, during the drafting process, the Office of the UN High Commissioner for Human Rights organised a Day of General Discussion (DGD) 'on a draft general comment on article 15 of the ICESCR: on the right to enjoy the benefits of scientific progress.'[16]

Almost a hundred people participated in the DGD, with twenty written contributions from various organisations, national and international, and experts, as well as *in situ* interventions of fifteen experts. Among these were Konstantinos Tararas (UNESCO), Jessica Wyndham (American Association for the Advancement of Science, AAAS), Lissa Bettzieche (German Institute for Human Rights), Prof. Brian Gran (Case Western Reserve University), Mylène Bidault (Observatory of Diversity and Cultural Rights), Prof. Helle Porsdam (UNESCO Chair in Cultural Rights, University of Copenhagen), Prof. Yvonne Donders (University of Amsterdam), Prof. Andrea Boggio (Bryant University), Prof. Lea Shaver, Gonzalo Remiro (Spanish Ministry of Science) and a few others.

The process of drafting this General Comment was possible thanks to the collaboration of various entities,[17] and benefited from the seminars, meetings, debates, and round tables organised for this purpose on several continents by various organisations, among which we are especially grateful to the Italian association Luca Coscioni, the AAAS, the UNESCO Chair on Cultural Rights at the University of Copenhagen and the German Institute for Human Rights. According to the UN:

> a general comment is a treaty body's interpretation of human rights treaty provisions, thematic issues or its methods of work. General comments often seek to clarify the reporting duties of State parties with respect to certain provisions and suggest approaches to implementing treaty provisions.[18]

Each treaty body (each committee) may adopt General Comments covering its interpretation both on substantive matters of the rights recognised or protected in each treaty, and on aspects relating to the accountability of States Parties and/or to institutional aspects.

These General Comments cannot in principle create new obligations for States Parties, but they can update the interpretation of the generic obligations contained in the treaty. This may have a remarkable importance in the progressive development of international human rights law. Though not *prima facie* legally binding, General Comments can therefore enjoy a great deal of legal authority.[19]

CONTENT OF THE GENERAL COMMENT

It is not my intention here to summarise the contents of General Comment No. 25, but it may be appropriate to identify some of its hallmarks.

The General Comment has seven parts. These are: an introductory part, followed by a second part on the normative content; a third part on elements and limitations; a fourth one on obligations; a fifth one on special aspects; a sixth one on international cooperation; and a seventh one on the national implementation of the law. I cannot in this brief overview go into more detail with each of these parts but will merely mention some of the most important points or issues. I hope that this will be enough to arouse sufficient interest in the reader to look up General Comment No. 25 in its full version.

2.1 On the Name

The General Comment did not directly adopt the wording 'Right to Science.' It merely states that:

> UNESCO, declarations made at international conferences and summits, the Special Rapporteur on cultural rights, and eminent scientific organizations and

publications have upheld the 'human right to science,' referring to all the rights, entitlements and obligations related to science.[20]

Other authors previously proposed other wordings. Lea Shaver had proposed the wording, 'the Right to Science and Culture,' to include the contents related to art, culture and science of Article 27, for example.[21] Early on, in 1970, B. Boutros-Ghali[22] had suggested that the right be called 'the right of access to knowledge' – a name that was later picked up by the Information Society Project of Yale Law School.[23]

The Committee, however, opted for a different wording in this General Comment. Instead of referring to 'the right to enjoy the benefits of scientific progress and its applications' (REBSP), it opted for 'the right to participate in and to enjoy the benefits of scientific progress and its applications' (RPEBSPA).[24] The Committee thereby wished to emphasise the double component of participation in science and enjoyment of its benefits, in fidelity to the Universal Declaration of Human Rights.

2.2 On the Normative Content and on the Concept of Science

The term 'normative content' refers to the question of what the law demands, and what rights and freedoms it recognises and protects. The normative content of the RPEBSPA includes access, enjoyment of the benefits of science, participation in that part of cultural life to which science belongs, as well as the freedom to do science, the protection of the moral and material interests of the authors, and actions for conservation, development and scientific dissemination.

As a part of the normative content of the RPEBSPA, the Committee was obliged to consider the definition of science for the purposes of this right and for the General Comment. To this end, reference was made to the authority of UNESCO, directly citing paragraphs of its 2017 Recommendation on Science and Scientific Researchers:[25]

> The word 'science' signifies the enterprise whereby humankind, acting individually or in small or large groups, makes an organized attempt, by means of the objective study of observed phenomena and its validation through sharing of findings and data and through peer review, to discover and master the chain of causalities, relations or interactions; brings together in a coordinated form subsystems of knowledge by means of systematic reflection and conceptualization; and thereby furnishes itself with the opportunity of using, to its own advantage, understanding of the processes and phenomena occurring in nature and society (para. 1 (a) (i)).

UNESCO adds that "the term 'the sciences' signifies a complex of knowl-edge, fact and hypothesis, in which the theoretical element is capable of being validated in the short or long term, and to that extent includes the sciences concerned with social facts and phenomena" (para. 1 (a) (ii)).[26]

The Committee concluded that:

> science, which encompasses natural and social sciences, refers both to a process following a certain methodology ('doing science') and to the results of this process (knowledge and applications). Although protection and promotion as a cultural right may be claimed for other forms of knowledge, knowledge should be considered as science only if it is based on critical inquiry and is open to falsifiability and testability. Knowledge which is based solely on tradition, rev-elation or authority, without the possible contrast with reason and experience, or which is immune to any falsifiability or intersubjective verification, cannot be considered science.[27]

2.3. Scientific Freedom in the General Comment

In a work such as the present one, which focuses directly on the question of scientific freedom, it is perhaps pertinent to make a very special mention of the thirteenth paragraph of the General Comment that was devoted to this question, as well as to other paragraphs in which scientific freedom is indicated.

Based on Article 15(3) of the Covenant, which establishes a specific duty for States to 'respect the freedom indispensable for scientific research,' the General Comment reaffirms in a very solemn manner the principle that '[i]n order to flourish and develop, science requires the robust protection of free-dom of research.'[28]

Beyond reductionist readings, Article 15(3) could be, should be and was in fact read by the Committee as containing legal elements conducive to both positive and negative obligations. For this reason, the Committee establishes a list of consequences or derivatives related to scientific freedom that in a very linked manner refer to both positive and negative aspects, in the form of freedom, liberties and entitlements:

> protection of researchers from undue influence on their independent judgment; the possibility for researchers to set up autonomous research institutions and to define the aims and objectives of the research and the methods to be adopted; the freedom of researchers to freely and openly question the ethical value of certain projects and the right to withdraw from those projects if their conscience so dictates; the freedom of researchers to cooperate with other researchers, both nationally and internationally; and the sharing of scientific data and analysis with policymakers, and with the public wherever possible.[29]

Pay attention, as an example, to the first sentence: 'protection of researchers from undue influence on their independent judgment.' Here, there is not just an acknowledgement of a right for researchers not to be 'unduly influenced,' and therefore a mere obligation for the state not to interfere; there is also a recognition of a right that implies a duty for the state to protect it in an active and effective manner. This example shows that a formalistic differentiation of positive and negative freedoms and positive and negative obligations is not always possible. Or, to use the language of the General Comment, the obligation to respect, to protect and to fulfill are rarely strict compartments; they are different approaches to a complex treatment of any right. That is why, when the issue of typologies of obligations is reached later in the General Comment, the question of scientific freedom is treated in a more detailed manner in the paragraph devoted to the 'obligation to fulfill':

This includes approving policies and regulations that foster scientific research, allocating appropriate resources in budgets and generally creating an enabling and participatory environment for the conservation, the development and the diffusion of science and technology. This implies, inter alia, protection and promotion of academic and scientific freedom, including freedom of expression and freedom to seek, receive and impart scientific information, freedom of association and freedom of movement; guarantees of equal access and participation of all public and private actors; and capacity-building and education.[30]

The treatment of freedom in the previously mentioned paragraph 13 is linked to different issues, particularly to the ideas of conscience and cooperation, which speak volumes about a view of the nature of science as a universal enterprise in which freedom, individual conscience and cooperation, collaboration or communication among colleagues, are necessary. All the three points of this triangle are interconnected.

A final point needs to be mentioned here: the relation between freedom and limitations. This point will be dealt with later; here it is just necessary to admit that scientific freedom is subject—as any other right—to limitations within the framework of the Universal Declaration and the Covenants.

A topic of close relevance is academic freedom. Though only touched upon in a very slight and indirect manner in this General Comment ('this implies, inter alia, protection and promotion of academic and scientific freedom'),[31] due to its intersection with other issues and mandates, it is a topic which deserves to be treated with the full attention shown to the elements of this General Comment.[32]

2.4. Elements of the Right

Following the path paved by other General Comments, the main elements of the RPEBSPA were directly identified and studied. Elements are those essential characteristics that compose and explain the right. Special attention was paid to these elements for this right to be understood and applied, for example: availability, accessibility, quality and acceptability.[33]

2.5 Limitations

Limitations are a key issue that must be studied in relation to any right. Its practical consequences are as delicate as they are necessary in every case, but perhaps in relation to the question of science they require special attention and finesse. In any event, as it could not be otherwise, the Committee relies on Article 4 of the Covenant, which recognises that rights may be subject to 'only to such limitations as are determined by law only in so far as this may be compatible with the nature of these rights and solely for the purpose of promoting the general welfare in a democratic society.'

2.6. Triple Typology

The scheme of triple typology of obligations is perfectly applicable to this right: to respect (not to violate), to protect (to prevent third parties from violating it) and to fulfil (to take measures to allow their enjoyment).

In relation to *the obligation to respect*, the State must refrain from 'interfering directly or indirectly in the enjoyment of this right. Examples of the obligation to respect are: eliminating barriers to accessing quality science education and to the pursuit of scientific careers; refraining from disinformation, disparagement or deliberate misinformation intended to erode citizen understanding of and respect for science and scientific research; eliminating censorship or arbitrary limitations on access to the Internet, which undermines access to and dissemination of scientific knowledge; and refraining from imposing, or eliminating, obstacles to international collaboration among scientists, unless such restrictions can be justified in accordance with article 4 of the Covenant.'[34]

With relation to *the obligation to protect*, the state should consider, among other things, 'ensuring that scientific associations, universities, laboratories and other non-State actors do not apply discriminatory criteria; protecting people from participating in research or tests that contravene the applicable ethical standards for responsible research and guaranteeing their free, prior and informed consent; ensuring that private persons and entities do not disseminate false or misleading scientific information; and ensuring that private investment in scientific institutions is not used to unduly influence the orientation of research or to restrict the scientific freedom of researchers.'[35]

In relation to *the duty to fulfil*, States should 'adopt legislative, administrative, budgetary and other measures and establish effective remedies aimed at the full enjoyment of the right to participate in and to enjoy the benefits of scientific progress and its applications. They include education policies, grants, participation tools, dissemination, providing access to the Internet and other sources of knowledge, participation in international cooperation programmes and ensuring appropriate financing.'[36]

2.7 Core Obligations

Core obligations are those duties that must be fulfiled as a matter of priority. Should these basic obligations not be fulfilled, they require a formal and detailed explanation from the State demonstrating that it has made every effort to fulfil them, using to the maximum its available resources, both individually and through international assistance and cooperation:

Among these core obligations, the Committee listed some related to:

> eliminating laws, policies and practices that unjustifiably limit access by individuals or particular groups to science, scientific knowledge and its applications; identifying and eliminating any law, policy, practice, prejudice or stereotype that undermines women's and girls' participation in scientific and technological areas; developing participatory frameworks; ensuring access to the basic education and skills; ensuring access to those applications of scientific progress that are critical to the enjoyment of the right to health and other economic, social and cultural rights; aligning government policies and programmes with the best available, generally accepted scientific evidence; promoting accurate scientific information and refraining from disinformation; protecting people from the harmful consequences of false, misleading and pseudoscience-based practices; and fostering the development of international contacts and cooperation.[37]

EFFECTS AND CONCLUSION

The adoption of a General Comment is a lengthy process. In our case, General Comment No. 25 took about six years to finish—a time surely longer than normal and desirable. Internal reasons within the Committee (related to other general comments that were being worked on previously and for that reason had been given priority), as well as reasons related to the content and ambition of the proposal can perhaps explain the length of this process. There are no doubt other general comments that have taken longer to finish, but perhaps the ideal period of time should be close to half of this, about three years approximately.

There are general comments that go down in the history of human rights law without much impact; let's say that their effects are reduced and even in some cases end up arousing only the interest of academics and experts. There are other general comments which have had a very notable influence on the development of human rights and on the enjoyment in real life of those rights. As an example of this last type of extremely important general comments we could mention General Comment No. 15 on water,[38] which contributed to the development of dynamics that affected not only the evolution of the right to water in international law and in the practice of international organisations,[39] but also in national legislation, and even inspired constitutional articles.[40]

As of the delivery of this writing, only two and a half years have passed since the adoption of General Comment No. 25, so we could hide behind the lack of time and sufficient perspective to avoid making any judgment on whether this particular general comment has had important practical effects. But the truth is that this time has not been two and a half normal years. The international community suffered a pandemic that has changed the way we contemplate some of the elements mentioned in the General Comment. When the WHO declared COVID-19 a pandemic and when states began to adopt exceptional measures in the face of an exceptional situation, the international community had an instrument in the shape of the General Comment that could give some guidance.

The General Comment has been studied in the academic field in these months, has served international organisations, has inspired public plans and programs,[41] and has even inspired constitutional proposals.[42] Its impact has been undeniable:

> UNESCO was the first international organization to incorporate the Human Right to Water focus into its mandate.[43] This happened after the approval of the General Comment on Water mentioned above. The same thing happened with regard to the adoption of *UNESCO brief on the right to science and COVID-19*[44] that refers, in a very direct manner, to the General Comment as a legal and conceptual basis for UNESCO's approach.[45] In addition, UNESCO has initiated courses,[46] events,[47] and publications[48] with a view to the conceptual development and practical dissemination of the human right to science.
>
> The OEI (Organización de Estados Iberoamericanos) was the first international organisation that politically supported the process of creating a General Comment on Science with a common declaration of heads of governments.[49] Once the General Comment had been adopted, it organised several activities in order to develop the human right to science in its mandate.[50]

The adoption of the General Comment facilitated the legal framework to introduce issues related to the human right to science in the reporting system of the treaty bodies, especially, but not only, before the CESCR. It should facilitate and draw attention to the presence of that right in the UPR system in the future, in order that its normative content can be consolidated and developed both legally and in practice.

From an international law perspective, the adoption of the General Comment marked a very significant milestone, but as I have explained, this was neither the beginning nor the end of the journey. The collaborative efforts by academia, civil society, scientists, public institutions, states and OIs should both implement and develop the important content of this right and open new roads for future, innovative approaches to face new challenges.

We can therefore affirm that GC 25 came at a time when the international community needed it, and that it has had a certain impact on the way we can approach the relationship between science and human rights, democracy and participation, both locally and internationally.

Chapter 8

Litigating the Right to Science before the CESCR: The View from the Trenches

Cesare P. R. Romano

UN human rights treaty bodies, such as the Committee on Economic, Social and Cultural Rights (CESCR), carry out three basic functions concerning the treaty over which they have been given mandate. First, they monitor its implementation by considering periodic reports submitted by States Party and issue 'concluding observations.' Second, they clarify its content by publishing occasionally 'general comments.' Lastly, they can consider particular cases of alleged violations of human rights committed by a State Party and submitted by individuals as an 'individual communication,' and issue 'views' on the matter.

The 'individual communications' process, to use the UN jargon, is optional. States that have ratified a given UN core human rights treaty, such as the International Covenant on Economic, Social and Cultural Rights (ICESCR), are not subject to the individual communications process unless they have given specific consent to that. In the case of the CESCR, specific consent is given through the ratification of the Optional Protocol to the ICESCR.[1] The UN General Assembly adopted it only in 2008, about thirty-two years after the ICESCR entered into force (1976), betraying a perduring reluctance of States to subject themselves to adjudicative processes when it comes to economic, social and cultural rights. It took five more years for the Optional Protocol to enter into force, on 3 May 2013, upon deposit of the tenth instrument of ratification.[2] The number of States that have ratified the Optional Protocol is still low. As of September 2022, out of 171 States which have ratified the ICESCR,[3] only 26 have also ratified the Optional Protocol.[4] Amongst them, the only major States to have done so are Argentina, Belgium, France, Finland, Italy, Spain and Venezuela.

Since the Optional Protocol entered into force, the Committee has received about ninety individual communications.[5] The overwhelming majority of them were brought against Spain, almost all on a single issue: the alleged violation of the right to housing. The others involve only a handful of States, including Argentina (1), Ecuador (4), France (2), Italy (1), Luxembourg (1), Portugal (1) and Uruguay (1). So far, the only two communications to have claimed a violation of Article 15 of the ICESCR are *A.M.B. v Ecuador,*[6] and *S.C. and G.P. v. Italy.*[7] However, *A.M.B.* alleged violation of the right to participate in cultural life (Art. 15.1.a), and, in the specific case, the right to participate in sport activities, while only *S.C. and G.P.* alleged, *inter alia*, a violation of the right to science (i.e. Art. 15.1.b, Art. 15.2 and 15.3). I had the honor to be one of those preparing and submitting both communications to the CESCR.

S.C. and G.P are a heterosexual couple who sought assistance to conceive a child from a private clinic in Italy specialising in assisted reproductive technology. In Italy, assisted reproductive technology is regulated by Law 40/2004.[8] In its original form, before it was shriveled by a series of decisions of the Italian Constitutional Court[9] and of the European Court of Human Rights,[10] Law 40/2004 was far-reaching and fundamentalist. It limited the number of embryos that can be produced during an in vitro fertilisation cycle to three, prohibited pre-implantation genetic diagnosis, mandated the simul-taneous transfer into the uterus of all embryos, regardless of their viability or genetic disorders, prohibited any clinical and experimental research on human embryos, and prohibited their cryopreservation.[11]

In early 2008, the victims requested the clinic to produce at least six embryos, to screen them with pre-implantation genetic diagnosis to identify possible genetic disorders, and not to transfer into S.C.'s uterus those embryos with genetic disorders.[12] The clinic declined the requests because incompat-ible with Law 40/2004.[13] The couple sued, challenging the constitutionality of the relevant provisions of Law 40/2004. The judge referred the matter to the Constitutional Court, and, pending the decision, ordered the clinic to pro-ceed with the fertilisation of three embryos, the maximum allowed by Law 40/2004, but also to carry out pre-implantation genetic diagnosis.[14]

The screening revealed that all three were affected by hereditary multiple osteochondromas (HMO), a genetic disorder the parents were suspected to carry because of their family history. Hereditary multiple osteochondromas, also known as hereditary multiple exostoses, is a highly transferable genetic disorder, with a high penetrance.[15] It causes people to develop multiple benign (noncancerous) bone tumors called osteochondromas, causing bone deformities that, depending on number, size and placement, might also be visible on the outside, and it is most active in children and adolescents.[16]

On 8 May 2009, the Constitutional Court declared Articles 14.2 (limiting the creation of a maximum of three embryos per in vitro fertilisation cycle, and imposing the simultaneous transfer all of them in utero) and 14.3 (which does not provide that the transfer of the embryos should be made without prejudice to the health of the woman) of Law 40/2004 unconstitutional.[17] Once these legal obstacles had been removed, in October 2009, the victims tried a second in vitro fertilisation cycle, at the same clinic.[18] This time, ten embryos were produced.[19] For technical reasons, pre-implantation genetic diagnosis could only be carried out on six of the ten embryos.[20] Of the six embryos screened, five had the genetic mutation causing hereditary multiple osteochondromas. The one that did not was deemed to have a low chance of nesting once implanted.[21] S.C. declined to have it transferred into her uterus. However, the clinic insisted that, according to their understanding of Law 40/2004, consent to transfer embryos into the uterus could only be revoked before fertilisation has taken place. Law 40/2004 is silent as to whether consent can be withdrawn after fertilisation. However, an interpretation that considers the spirit of the law (i.e., the inviolability of the embryo) suggests it cannot. Facing the possibility of a lawsuit if she declined to have the embryo implanted, S.C. agreed.[22] Unfortunately, but not unexpectedly, she suffered a miscarriage. The other nine embryos were put in cryopreservation. The authors requested that the clinic surrender them because they intended to donate them to scientific research.[23] However, the clinic refused because article 13 of Law 40/2004 prohibits research on embryos.[24]

On 30 March 2012, the victims filed a lawsuit against the clinic and the Italian Government, requesting the court to order the clinic to surrender the embryos, and to determine the validity of S.C.'s decision not to have the embryos transferred into her uterus.[25] On 7 December 2012, the court referred the matter to the Constitutional Court, asking to determine the compatibility of Articles 6.3 (regarding the revocation of the consent before fertilisation) and 13 (regarding the prohibition of research on embryos) of Law 40/2004 with the Constitution.[26] Almost four years afterwards, the Constitutional Court declared the request inadmissible because moot, since S.C. had agreed to have the embryo transferred into her uterus, while the question of future consent would be speculative.[27] However, recognising that the case had multiple ethical and juridical implications, on a matter that divides jurists, scientists and society, the Constitutional Court called on the legislator to strike a better balance between the rights of the embryo and the right to enjoy the benefits of scientific progress and its applications.[28] Regrettably, this has not yet happened. The Italian legislator has not yet amended Law 40/2004 to reflect the several decisions that the Constitutional Court and human rights courts and bodies have adopted over the years, and, given the current state of Italian politics, it is unlikely this will happen in the near future.

Having exhausted all domestic remedies, the victims brought their case to the CESCR, alleging violation of Articles 10 (i.e. rights of the family and reproductive rights), 12 (i.e., right to health) and 15 (1) (b) (i.e. right to science), 15(2) (i.e. duty of the State to conserve, develop and the diffuse science) and 15(3) (i.e. freedom of research) of the Covenant.[29] The Committee easily found Italy in violation of Article 12 because S.C.'s consent to the transfer in utero of the fertilised embryo was not valid, since it was obtained under threat of legal action, and because Law 40/2004 should allow withdrawal of consent to transfer in utero even after fertilisation has occurred.[30] It declined to decide on the question of the violation of Article 10 because the substance of that claim had already been addressed by its analysis of Article 12.[31] However, as to the Article 15 claims, it found them inadmissible.[32]

The arguments presented and the reasoning of the Committee shed much light on both the limitations and potential of the communications procedure to advance the right to science. The victims argued that Italy had violated Article 15(1)(b) (right to benefit from progress in science and technology) of the Covenant because, by prohibiting research on embryos, Law 40/2004 interferes with scientific progress and slows down the search for a cure for various diseases.[33] Also, by preventing them from donating to research the embryos affected by the genetic disorder, Law 40/2004 was claimed to have violated S.C.'s and G.P's right to participate in scientific research.[34]

The victims pointed out the arbitrariness of many aspects of Law 40/2004. First, the notion of 'embryo' in Law 40/2004 is not scientific. While scientific research deems an embryo to exist several days after the fertilisation, according to Law 40/2004 the embryo exists from the moment of fertilisation.[35] Second, although Italy prohibits research on embryos because they are deemed human life that must be protected, in Italy scientists can do research with stem cell lines that have been created abroad through the destruction of embryos.[36] Third, the cumulative effect of the prohibition of research on embryos and the declaration of unconstitutionality of the prohibition of pre-implantation genetic diagnosis has created an absurd situation whereby embryos affected by a genetic disorder are not transferred in utero, cannot be used for scientific research, and cannot be disposed of. The only option is to keep them in cryopreservation *sine die*.[37]

S.C. was an asymptomatic carrier of hereditary multiple osteochondromas. Given the high transmissibility and penetrance of the disorder, without a cure, her chances of conceiving a healthy child were slim. It was argued that, by preventing the donation of affected embryos to research aimed at finding a cure, Italy violated not only the donors' right to benefit from scientific research, but also their right to participate in scientific research; a right that can be found if Article 15(1)(b) is read in conjunction with Article 27 of the Universal Declaration on Human Rights.[38] Also, it was argued that Law 40/2004 prevents Italy from fulfilling its duty to develop and disseminate scientific developments (Art. 15(2) of the Covenant).[39] Prohibiting research on human embryos makes it harder for scientists to realise the potential of research, for instance, on stem cells, and hinders the spreading of scientific knowledge and applications within the scientific community and society at large. Finally, Italy was accused of failing to respect the freedom indispensable for scientific research and creative activity (Art. 15 (3) of the Covenant) by blocking the research on embryos without a legitimate purpose, since, in the present case, there was no other legitimate right to be protected as the embryos concerned will be forever in a frozen limbo.[40]

The Committee rejected the Article 15 claims as inadmissible for lack of standing of the victims. Article 2 of the Optional Protocol restricts the *locus standi* for submitting communications only to those who have suffered a cognisable prejudice because of the violation of the rights protected in the Covenant. 'The Committee may not examine a communication in abstracto: it may not assess whether an action or an omission of a State party is compatible with the Covenant, unless such action or omission has affected the author.'[41] '[F]or the Committee to enter into the merits of a communication, it is necessary for the facts and the claims presented to show, at least *prima facie*, that the authors might be actual or potential victims of the violation of a right enshrined in the Covenant.'[42]

Simplifying the various complex Article 15 arguments made by the victims, the Committee reformulated them into three claims, all of which it rejected. The first is that, by preventing the donation of the embryos to science, research on hereditary multiple osteochondromas would be hindered. However, because the communication 'does not provide any minimum level of evidence that the donation of these specific embryos would produce any concrete benefit for the authors in relation to hereditary multiple osteochondromas, [and i]t is not even clear at all that the embryos would be used in research on this disease . . . , the argument about the benefits for the authors remains speculative,' and must be rejected.[43] 'Had the authors provided sufficient evidence that there was a probable, or at least a reasonable, link between the donation of these specific embryos and the development of better treatments for the disease or the reduction of the probability of its hereditary transmission, that would benefit them personally, their claim would have been admissible.'[44] The second claim the Committee read in the communication was that by prohibiting research on embryos, Law 40/2004 violated S.C. and G.P.'s right to participate in scientific research. The Committee rejected it because the authors did not 'substantiate in any meaningful manner that a donation of an embryo is really a form of participation in scientific research.'[45] The third claim, that the restrictions imposed by Law 40/2004 violate the obligation of States to respect the freedom indispensable for scientific research (Article 15(3) of the Covenant), was also rejected because the 'the authors have never claimed that they intended to perform any scientific research themselves, so in reality they are not claiming that they might be victims of a violation of their freedom of research.'[46]

It is a pity that the Committee did not take the occasion offered by this case to elaborate more on the normative content of the right to science. Perhaps it felt hesitant doing so. As the Committee considered *S.C. and G.P. v. Italy*, it was also discussing and drafting General Comment 25 on the Right to Benefit from Progress in Science and Technology,[47] and, it was struggling mightily with coming to a consensus on its content. Perhaps it was hampered by a fundamental misunderstanding of how biomedical scientific research works. After all, amongst the committee members there was no one with a meaningful background in natural sciences.

Of course, in the future, other cases claiming violations of the right to science or some of its discreet components might reach the Committee. However, one must wonder whether with this decision the Committee has shut the door too firmly to similar cases. Indeed, if one were to apply the tests set by the Committee in *S.C. and G.P. v. Italy* to possible future claims, it is hard to imagine what a successful case might look like.

First, one would have to show that that the cells, tissues or organs that are prevented from being donated will be used in research. Anyone working in a laboratory knows that that cannot be guaranteed, and even if it was, then there is the question of what evidence would suffice for the Committee. Perhaps a statement to that effect by the laboratory? One can only speculate. Second, one would need to be able to prove that there is 'a probable, or at least a reasonable, link between the donation and the development of better treatments for the disease or the reduction of the probability of its hereditary transmission.'[48] That is much harder. Anyone with a minimum of scientific training understands that it is extremely difficult to tell *ex ante* whether the donation of specific cells, tissues or organs to research will actually produce concrete benefits for the donor, not even if the evidentiary threshold is only 'probability' or 'reasonable expectation.' The only scenario where one could show 'probability' or 'reasonable expectation' is the one of the donation of organs for transplants. Third, one needs to show that the author of the communication would personally benefit from the discovery of the treatment. Scientific research and the development of therapies can take years or generations to produce results. How many donors in medical history have benefited themselves from scientific discoveries made thanks to cells, tissues or organs they donated? That is simply not the way biomedical research works. Moreover, according to the reasoning in *S.C. and G.P. v. Italy*, only scientists can advance claims of violation of the freedom of scientific research, and only if their own personal freedom is violated and their own research is hindered. Unless donors can demonstrate persuasively that donation of cells, tissues or organs is a form of participation in scientific research, they cannot claim a violation of the freedom of scientific research. In other words, according to the Committee, Henrietta Lacks, the woman whose cancer cells are the source of the so-called HeLa cell line, the first immortalised human cell line and one of the most important cell lines in medical research,[49] never 'participated' in scientific research.

Incidentally, one could wonder how the Committee would consider those who donate money to scientific research. Would that qualify as 'participation in research' for the purposes of the application of Article 15? Would they have a valid claim when the research to which they donated is hindered?

We will have to wait until a similar case reaches the Committee to answer all those pregnant questions. However, considering the small number of States that have ratified the Optional Protocol, its admissibility criteria, which require finding a victim who has demonstrably suffered a prejudice and has exhausted all available domestic remedies, and a certain understandable reluctance of the Committee to wade into difficult, cutting edge, questions, it might be years, if not decades, before the Committee will have another chance.

Chapter 9

Defending Science, Knowledge and Facts: The UN And Scientific Freedom of Expression

Malene Nielsen and Carsten Staur

COVID-19 has led governments across the world to impose new and intrusive restrictions on civic space. The enjoyment of scientific freedom is one of those civic spaces that is currently shrinking at a worrisome speed, and the right to freedom of expression for scientific researchers is increasingly questioned, if not outright violated across the world. While some scientists may have become global heroes, taking part in the development of vaccines against the coronavirus, many others have faced various attacks and/or intimidations, offline as well as online. Several researchers have experienced restrictions in their COVID-19-related scientific work, including attempts to restrict research and the flow of information about the virus. Many researchers who have been voicing criticism of governmental responses to COVID-19 have faced disproportionate restrictions by governments.[1]

The accelerated move to online public spaces, not least because of COVID-19 lockdowns, has amplified the digital flow of scientific information, but it has also increased the digital risks. Like other professionals such as journalists and artists, whose work also increasingly depends on communication via online platforms, scientific researchers face new challenges in the form of online harassment, the rise of digital authoritarianism and the rapid spread of disinformation on the Internet.[2]

These attacks have a chilling effect on the working environment of scientific researchers, undermining the scientific freedoms indispensable for the conduct of proper research. While knowledge, insights and data on these worrisome trends are currently limited, the United Nations have been already sounding the alarm. The UN Secretary-General has stated that it is time 'to end the infodemic' and the 'war on science,' calling upon UN senior management and staff to take steps according to their respective mandates, individually or jointly, to protect groups at particular risk.[3] In addition, the Assistant Director-General of the UNESCO in charge of Social and Human Sciences has called upon government officials, scientists, and UN entities to cooperate, with the objective of building an 'ecosystem for free thought and research' in compliance with the UNESCO 2017 Recommendation on Science and Scientific Researchers.[4]

The digital threats have also been addressed by the UN Deputy High Commissioner for Human Rights (OHCHR) who has recommended to the UN Human Rights Council (HRC) that accountability and transparency 'should guide state regulation of new technologies and private sector behavior.' To that end, the OHCHR is developing UN system-wide guidance, complementary to the existing UN Guiding Principles on Business and Human Rights. A recent UN stocktaking project points to the need for the Universal Periodic Review (UPR) of HRC to use the UNGPs systematically to address and assess States' performance, including in relation to law and policy on tech companies and human rights.[5]

This concurs with the findings of a Danish initiative 'Critical Voices,' which highlights the need to further use the UPR to hold Member States accountable for their obligations to reinforce scientific freedom; and to ensure a safe and enabling working environment for scientific researchers.[6]

SCIENTIFIC FREEDOM BASED ON HUMAN RIGHTS

The right to freedom of expression is enshrined in Article 19 of both the Universal Declaration of Human Rights (UDHR) and the International Covenant on Civil and Political Rights (ICCPR). Both state that everyone is entitled to seek, receive and impart information and ideas of all kinds regardless of frontiers through any media. This involves the right to information and provides a wide understanding of expressions to be protected—whether the expression is considered true, false, offensive or enlightened, keeping in mind that scientific advice may change as new data and understanding become available. It also anticipates the continued development of media, including new technologies for the dissemination of scientific knowledge.[7]

States are required to respect and protect the freedom indispensable for the conduct of scientific research, as explained in the General Comment No. 25 concerning Article 15 of the International Covenant on Economic, Social and Cultural Rights (ICESCR). States must take steps to eliminate laws, policies and practices that unjustifiably limit the freedom of scientific research.

At the same time, States must promote accurate scientific information and refrain from disinformation that deliberately misinforms the public or erodes public trust and respect for science and scientific research. Scientific researchers should be protected from undue influence and should have the freedom to share scientific data and analyses.[8]

Some restrictions on scientific freedom may be legitimate during a state of emergency, but such measures must comply with international standards. The emergency must amount to a public emergency that threatens the life of the nation, and the government must have officially proclaimed the state of emergency and communicated the end-date of the derogation. The Human Rights Committee's statement in connection with the COVID-19 pandemic specifically underlines that 'freedom of expression and access to information and a civic space where a public debate can be held constitute important safeguards for ensuring that States Parties resorting to emergency powers in connection with the COVID-19 pandemic comply with their obligations.'[9]

SCIENTIFIC RESEARCHERS AS A GROUP AT RISK

Member States have the obligation to ensure an enabling working environment for scientific researchers. This duty is elaborated in the UNESCO 2017 Recommendation on Science and Scientific Researchers, which stipulates that Member States should encourage conditions that deliver high-quality science and ensure that scientific researchers can 'pursue, expound and defend the scientific truth as they see it.' This may include the freedom of researchers to freely and openly question or withdraw from governmental projects. Member States are obliged to recognise that it is 'wholly legitimate, and indeed desirable, that scientific researchers should associate to protect and promote their individual and collective interests, in bodies such as trade unions, professional associations.' The Recommendation specifies that, 'in all cases, where it is necessary to protect the rights of scientific researchers, these organizations should have the right to support the justified claims of such researchers.'[10]

According to a 2022 survey in *Science*, scientists who have publicly advocated hotly debated positions on COVID-19, particularly on social media and in news stories, have been more harassed than other scientists who were less visible in the public debate.[11] Scientific researchers in the field of climate change have faced similar threats and attacks for decades. Such attacks on science and scientists have been highlighted by experts as a common feature of the rise of authoritarianism, stressing that authoritarians and extremists use common tactics to silence researchers.[12]

The UN system is well placed to recognise this worrisome pattern and may draw useful parallels from its wealth of information on the threats to the safety of journalists and media freedom. In a recent report, the UN Special Rapporteur on freedom of expression stated that, 'whether online or offline, the objective of those who threaten journalists remains the same: to chill public interest reporting by increasing the risks faced by journalists.'[13] The recent UN system-wide Guidance Note—prepared to pursue the UN Secretary-General's 'Call to Action for Human Rights' – does mention journalists, but not scientific researchers, as being part of the group of civic space actors that requires special attention across the UN system. UNESCO is particularly well placed to encourage the UN system to include scientific researchers in its efforts to 'be aware and raise awareness of legislative, institutional and policy contexts and groups at risk,' and 'ensuring increased consistency across the system,' as required in the Guidance Note.[14]

The potential of raising awareness within the UN system through joint efforts of data collection should not be underestimated. Knowledge-building regarding legislative issues on media freedom and the safety of journalists has laid the foundations for UN entities—Paris, Geneva, New York, UN Country Teams—to respond to challenges in a systematic and coordinated manner. The specific insights on the dynamics and root causes of the various types of attacks on journalists and on media freedom have also enabled the UN system to endorse a UN action plan, which outlines responsibilities and coordinate work among UN entities and other key stakeholders. Similar efforts could be introduced with respect to actions of the UN system to reinforce scientific freedom and the safety of scientific researchers.[15]

CORPORATE RESPONSIBILITY FOR THE SAFETY OF SCIENTIFIC RESEARCHERS

Concerns about the use and misuse of digital platforms, as well as the role of tech companies in mediating freedom of expression, have been identified as challenges to be addressed within the framework of the UN Guiding Principles on Business and Human Rights. These principles outline States' duties to protect against human rights abuses by tech companies. It is stressed that states may breach international human rights law obligations if they fail to take appropriate steps to prevent, investigate, punish and redress companies' abuse. Tech companies also have a responsibility to respect human rights and to conduct due diligence to identify and assess the human rights risks associated with their activities. They must establish clear policies in this respect and publish transparency reports on the risks they encounter and how these are addressed, and they must provide remedies in case of violations.[16]

Particularly relevant to scientific freedom and the safety of scientific researchers are the role and responsibilities of social media platforms—as enablers of the digital flow of scientific information, and as vectors of coordinated digital attacks against scientific researchers and viral disinformation that undermine science. More and more research points to the fact that social media platforms can do more to address the problem of online attacks. UNESCO has commissioned a research project that examines this issue from the perspective of journalism. It provides an assessment of big tech's response to online violence against women journalists but is much less focused on digital abuse and harassment of professionals in the field of science.[17]

However, recent surveys do indicate that scientific researchers are also confronted with the failure of tech companies to act against the online content and perpetrators involved in online violence. Anti-vaccination groups and individuals are among those who are driving the online attacks on scientific researchers. They rely on social media companies' algorithms to amplify their messages, knowing that fights as well as controversial tweets and posts drive visibility and, in turn, lead to more followers and greater impact. Experts also point to the fact that some of these actors are 'economically motivated' to drive online attacks. UNESCO's groundbreaking partnership with Netflix could provide a platform to discuss how companies, such as streamlining services, should respond when repressive governments request the removal of critical content.[18]

Institutions that employ, fund, govern or guide researchers are also responsible for upholding the rights of scientific researchers and ensuring their safety. A key question is how this responsibility can be put into practice? UNESCO's work on journalists is based on global research papers to guide operational responses to such questions. This is much less the case when it comes to the organisation's work to protect scientific researchers from attacks. In short, a common UN voice defending the rights of scientific researchers, journalists and artists may facilitate collective action by three professional groups, and hence the amplification of public voices who collectively defend the right to freedom of expression and online safety.[19]

Data, Transparency and Global Understanding of the Issues

The global data collection on the implementation of the UNESCO 2017 Recommendation falls within the purview of the UNESCO Conventions and Recommendations Committee (CRC) and its specific multi-stage procedure. This implies that all Member States are required to report on their implementation of the Recommendation every four years, based on a questionnaire from the UNESCO Secretariat. Each national report is considered an evidence-based self-assessment in which compliance should be substantiated by documentation and references, involving analysis that typically would be based on data collection and consultation to assess the impact of policy measures that have been taken. UNESCO is promoting the view that national reports on the Recommendation are important—and not only from an accountability perspective. If properly and regularly conducted, reporting can become a powerful and useful exercise that permits Member States to better understand their science system, identify patterns, derive actionable insights, and take measures to advance their science agenda.[20]

At present, this potential of national reporting to UNESCO remains largely untapped. Only 35 Member sStates submitted their reports to UNESCO for the first cycle of periodic reporting on the 2017 Recommendation. While civil society actors were consulted by UNESCO on the design of the questionnaire, only a minority of Member States have conducted consultations with their scientific communities prior to their reporting to UNESCO. This makes it difficult to justify very affirmative conclusions on how the Recommendation is implemented. Nevertheless, the Director-General's consolidated report does point out some important trends and issues, including by highlighting scientific freedom as one of the major issues. In particular, the report points out that in many countries, scientists have been jailed on charges suspected of being intended to silence their voice, tending to chill the atmosphere of freedom that is so central to conducting quality research. The report further stressed that UNESCO is aware of several existing, new and proposed legal measures that are susceptible to suppress or chill scientific freedom.[21]

The first consolidated report of UNESCO on the 2017 Recommendation clearly signals that UNESCO's efforts for greater transparency is a balancing act. The organisation is balancing between building its report on the submissions from Member States—plus a few civil society actors, when Member States decide to consult them—and other data sources, which are open to the public, and may help to shed further light and assess the global state of play of scientific freedom. Specific insights about the various types of attacks on scientific researchers, which are the foundations for the design of tailored policy responses, are almost absent in the official UNESCO documentation. The lack of data and transparency of social media companies, the so-called black box, is adding to the pile of unknowns. For instance, Member States are not provided with data on how social media platforms may enable some actors, be it anti-vaccination groups or others, to silence scientific researchers by way of systemic harassment.

UNESCO's intergovernmental decision-making process for following-up on the Recommendation is therefore handicapped by lack of official documentation, which would normally serve as the foundation for evidence-based decisions on policy responses and recommendations. So far, it is difficult to perceive UNESCO's report on the 2017 Recommendation as a tool which can create a realistic global understanding of the scope of the problems faced by science and scientific researchers. At the same time, and since a number of governments have used the COVID-19 pandemic to take intrusive measures to restrict freedom of expression for scientific researchers, external scrutiny and accountability are becoming extremely relevant. In 2013, a UNESCO expert group pointed out that, 'consideration could be given to explicit linkage between monitoring of the implementation of the Recommendation and UNESCO's inputs to the UPR, under the aegis of the Human Rights Council.' In the current context, there may be a renewed momentum for elaborating on this idea.[22]

Accountability, Diplomatic Pressure and Impact at Country Level

The Universal Periodic Review is a Member State–driven process, organised by the Office of the High Commissioner for Human Rights (OHCHR). In the UPR process, all countries can reflect on the human rights performances and experiences of each other and may also in this context present specific recommendations to the Member State under review. The government under review may – or may not—accept the recommendations provided by other Member States. If a government under review accepts a recommendation, this acceptance provides a—sometimes broadly framed, sometimes more specific—consensual approach to advancing human rights, often indicating a commitment of said government to undertake actions for improving its human rights situation in a certain area.

Referring to *accepted* UPR recommendations is thus not politically sensitive or controversial, as these have already been accepted by the Member State in question. Since the first UPR-cycle in 2007, Member States have received a total of around 80,000 recommendations from other Member States. The numbers vary from country to country. Of these 80,000 recommendations, around 60,000 recommendations have been accepted, underlining that all countries acknowledge that there is room for improvement in their protection of the human rights of their citizens. In accepting a recommendation, Member States may, of course, rely on their own interpretation of the scope of the recommendations in question. They may interpret it more narrowly, or more broadly, to be able to better live up to the commitments made.

For Member States presenting recommendations to other Member States, the political dimension may explain why some recommendations are formulated in very broad terms. As one observer has put it, 'states seem reluctant to criticize each other too harshly, to avoid jeopardizing their diplomatic relations.'[23] Evidence has shown that states generally prefer UPR recommendations to be: 1) specific and measurable; 2) achievable within a specific timeframe; and 3) able to consider the national context of the reviewed states. Based on the filtering of data done by the Danish Institute for Human Rights, it appears that a large number of recommendations are if not fully, then still more or less, in line with these standards.[24]

A research study indicates that the nature of bilateral relations—between the presenting and receiving country—play an important role, not only in the formulation of the recommendation, but also in the decision of the receiving country whether to accept it; and if accepted, in the country's effort to implement it. The study argues that since the accepted recommendations are endorsed by all states participating in the UPR, they become political commitments between countries with strong political implications. As a result, these commitments are more likely to be adhered to, 'when recommendations are delivered by a country with whom the reviewed state aims to maintain positive diplomatic relations.'[25]

Field research conducted during the first UPR cycle concurs with this observation, highlighting cases where receiving countries feel compelled to accept recommendations to please; or where presenting states may 'offer what were judged as 'easy' or friendly recommendations, thus crowding out recommendations addressing a state's deficiencies and/or misdeeds.' Such 'friendly' recommendations have been justified as a gesture of solidarity, or as a necessity for maintaining cordial relationships and protection in times of tense/conflictual relationships with other countries.[26]

Governments, national human rights institutions, and civil society organisations have developed tracking methodologies and tools to help monitor progress on the implementation of the various UPR recommendations. Further to this, Member States, stakeholders and UN entities can submit on a voluntary basis, so-called Mid-Terms Reports (MTRs) to reflect the current situation and the extent to which recommendations are being followed. MTRs can be an opportunity for national actors to detail the steps they are taking to implement accepted UPR recommendations, such as developing a national action plan. It has also paved the way for individual or collective submissions from CSOs, which have 'brought necessary independent perspectives.'[27] For the purpose of follow-up on the few, but strategically important, UPR recommendations that specifically relate to science and scientific freedom, the Danish initiative 'Critical Voices' has made the database available online for all interested stakeholders.[28]

SCIENTIFIC FREEDOM—CONNECTING
THE DOTS WITHIN THE UN SYSTEM

Around 2 per cent of all the accepted UPR recommendations—a little more than 1,200 accepted recommendations since 2007 – pertain to the freedom of expression, calling for, and often providing concrete guidance on, measures to improve the freedom of expression of certain groups at risk. Of these 1,200 recommendations on freedom of expression, the vast majority seek to improve freedom of expression for journalists. Only twenty refer to scientific researchers.[29]

'Critical Voices' points out that information and insights from civil society actors, as well as UN entities, constitute an important part of the knowledge base, assisting Member States' in their formulation of UPR recommendations. Of the twenty UPR recommendations relating to scientists, only a few appears fairly precise and pertinent, such as the recommendation to one country under review to 'Reconsider (. . .) legislation that restricts open and honest scientific research and that can serve to intimidate researchers.'[30] It is difficult not to see the lack of focus on scientific freedom—and the lack of precision in the UPR recommendations in this area—as anything but lack of awareness of the UNESCO related obligations of Member States in relation to science and scientific researchers, and a shortage of civil society capacities to monitor and report on specific issues in this respect. Another explanation could be that the right to enjoy the benefits of scientific progress is among the least developed human rights in the UN system.

In stark contrast to UNESCO's more proactive approach to freedom of expression for journalists, the organisation does not produce country-specific information about scientific freedom to the benefit of the UPR process. The UPR information provided by UNESCO merely points to the existence of the 2017 Recommendation. It does not—as is the case for information submitted by UNESCO regarding journalists—highlight or comment on national laws and policies in the country under review that may not comply with international standards on scientific freedom. Within the UPR framework, science and scientific freedom, including freedom of expression, is probably one of the areas to which Member States collectively have so far given the least attention.[31]

WHAT WILL STAKEHOLDERS MAKE OF EXISTING UN HUMAN RIGHTS TOOLS IN THE NEXT FIVE YEARS?

As the UN system is embarking on a new five-year UPR-cycle in 2022, there is a new opportunity to rally stakeholders for enhanced engagement, using the work of UPR more strategically to raise awareness about the issues of scientific freedom and the safety of scientific researchers.

A first, fundamental step would be to ensure that more information and documentation on the issues find its way to the UN system. Secondly, that the reported information is used strategically to increase transparency about how states fulfil obligations; and to hold states accountable on specific key issues that need to be addressed. To that end, important informal discussions have been launched—in Paris and Geneva—by diplomats, UN entities, and civil society organisations, based on the findings and recommendations of the report 'Critical Voices.' In addition, an international workshop was held on the 2022 World Press Freedom Day in Montevideo, Uruguay, with stakeholders from various affiliations, who have different individual interests, but are all keen to exchange views on one commonly shared concern: how best to address the challenges of freedom of expression for professionals at risk, such as scientific researchers, journalists and artists.[32]

From these discussions, an idea has evolved of creating a UPR coalition whose aim is to improve the enabling environment for advances in scientific freedom, through a more systematic use of the UPR process. At its best, such a coalition could play a vital role in building resilience to external pressures; securing consensus of what needs to be done; and driving collective action. Recent case studies on media coalitions point out that collective action through coalitions can give stakeholders leverage to bring about positive change at country level, in areas such as legislative reform, regulation and professionals' safety measures.[33]

Member States, civil society organisations, and the various entities of the UN system all have a fairly good understanding of the methods, tools and techniques of the UPR process. This is much less the case for scientific researchers and their professional organisation, whether at national or international level. The scientific community, in general, finds UNESCO's 2017 Recommendation highly relevant, and have expressed a clear interest in looking for ways to hold states accountable. The missing link is a stronger engagement of scientists in the new UPR cycle, including in submitting civil society information to OHCHR to raise awareness and feed the UN system with independent information on the specific situation of scientific freedom in the countries where the various stakeholders operate.

There is a clear need for the creation of a stronger UPR ecosystem regarding scientific freedom. A system where a few member state champions would take on a strategic role in introducing discussions on scientific freedom and asking questions, such as, e.g., what measures can x-country take to regulate big tech, based on international human rights standards, with the objective of protecting scientific researchers against online attacks. Further to this, state champions could play a crucial role in increasing the number and quality of UPR recommendations that address key issues of scientific freedom. In turn, UNESCO, as the UN entity with the specific mandate on science, would need to change its approach to the type of information it provides to the UPR process. General and one-size-fits-all information is not sufficient. UNESCO would have to take new bold steps towards country-specific information about scientific freedom in the country under review. In turn, this new step would facilitate the delicate exercise of Member States who would be responsible for formulating UPR recommendations that are specific, measurable, and achievable within a specific timeframe, and would be able to consider the national context of the reviewed country.

Overall, the UN Secretary-General has put the 'infodemic' and 'war on science' on the global radar. At the same time, it can be argued that the UN system's operational capacity to act collectively for better ecosystems of scientific freedom is heavily challenged, not least by a serious knowledge gap. Data on scientific freedom is limited in general. However, it can also be argued that the information and documentation that reaches the UN system is even more limited, bearing in mind that Member States, so far, constitute the main providers of the information to UNESCO. The lack of independent data sources that would inform UN documentation can be particularly troublesome, assuming many governments' unwillingness to expose their own weaknesses and authoritarian approaches.

The data gap makes it harder to design tailored policy recommendations and responses that adequately reflect the scope, nature and complexities of the issues in relation to scientific freedom. A first and fundamental step would be to mobilise stakeholders, demonstrating why it is important to report periodically to UNESCO and to inform the UPR process. A key factor for success would be to outline the different roles and propose actions to the various stakeholder groups. The UN system marked the tenth anniversary of the UN Plan of Action on the Safety of Journalists in November 2022, taking stock of achievements and lessons learnt, as well as devising new strategies that reflect the digital impact and invoke big techs' responsibility alongside states' duty. It may also be a timely moment—bearing in mind the UN priority to put an end to 'the war on science' – for reflecting and proposing coordinated UN action—a roadmap for the next five years—with the aim of reinforcing the ecosystem of scientific freedom, offline as well as online.

Appendix A: Systematic Review, Methods and Results

Whereas medical reviews often combine the numerical results of individual studies into an overall singular estimate of effect, the qualitative nature of our inquiry lent itself to a different approach to data gathering. Instead of statistically analysing effect sizes derived from interventional trials, our report aimed to identify key topics relevant to scientific freedom raised in the included studies. This step necessarily introduced some subjectivity and bias to our methodology. Although our decisions concerning which topics we considered as important enough to include were made against the background of extensive reading around the topic, the decisions were still essentially made by authorial judgement and fiat. Others conducting an analysis of our source material may have chosen other topics. However, this was unavoidable, and we carried out such mitigation as was feasible by having each selected topic screened by both authors. In no cases did a disagreement arise.

It should be noted that our search strategy, though successful in identifying many relevant sources, was not designed to be exhaustive. This was due to the extreme number of documents identified by more inclusive search syntaxes. Some of our early searches resulted in tens to hundreds of thousands of potentially relevant sources. Although it would be worthwhile to extend the analysis using such syntaxes, doing so was deemed excessive for present purposes and resources.

Thus, we restricted our search to three databases: PubMed (the digital version of the US National Library of Medicine), JSTOR (one of the largest scholarly databases) and Google Scholar. The Google Scholar search was restricted to the first ten pages of results.

The searches were carried out and yielded results as follows:

SEARCH RESULTS

Searches were carried out by SPM and MMS on 13 July 2019.

PubMed: search syntax: 'scientific freedom' OR (('scientific inquiry' OR 'scientific research') AND 'freedom') OR ('research freedom')

207 **Results**

Web of Science: search syntax: 'scientific freedom' OR (('scientific inquiry' OR 'scientific research') AND 'freedom') OR ('research freedom')

350 **Results**

Google Scholar: search syntax: 'scientific freedom' OR (('scientific inquiry' OR 'scientific research') AND 'freedom') OR ('research freedom')

100 **Results**

Duplicates: 60 duplicate articles were identified using the EndNote find duplicates function; an additional 41 more were identified during manual search, resulting in a total of 101 duplicates.

Thus, our search yielded 657–101 = 556 unique hits.

The relevance of these were judged against our inclusion criteria, which are set out below:

Inclusion criteria:

- The study concerns or mentions the concept of scientific freedom; OR,
- The study involves concepts related explicitly, or obviously implicitly related to, scientific freedom; AND,
- The study was published in English, German or Danish; AND,
- The study was published in a peer-reviewed journal listed on MEDLINE or JSTOR; or in the first ten pages of Google Scholar.

Exclusion criteria:

- The study mentions scientific freedom only once or several times in passing, and/or does not engage in reflection, analysis or commentary on the concept as such; OR,
- The study does not engage with concepts immediately relevant to or directly linked to scientific freedom or its main derivative concepts; OR,
- The study was published in a language other than English, German or Danish; OR,
- The study was not published in a peer-reviewed journal listed on MEDLINE or JSTOR; nor in the first ten pages of Google Scholar.

An initial screening by study title resulted in the exclusion of 314 studies. Another 96 studies were excluded based on their abstracts. Finally, 66 studies were excluded based on their full text. This yielded 80 studies fitting inclusion criteria.

This step of the analysis is depicted visually below using the 2009 updated PRISMA flowchart for systematic reviews:

The same protocol described above was repeated on 29 October 2022 by SPM & MMS. This resulted in the inclusion of four additional records.

The 84 included studies discussed a wide range of concepts relevant to scientific freedom. Based on our search, we identified 20 distinct themes within the included studies. These themes, as well as the number of times they were mentioned, are illustrated in Table One below. Note that, due to a loss of data, the figure is based on 74 of the 84 included studies.

 PRISMA 2009 Flow Diagram

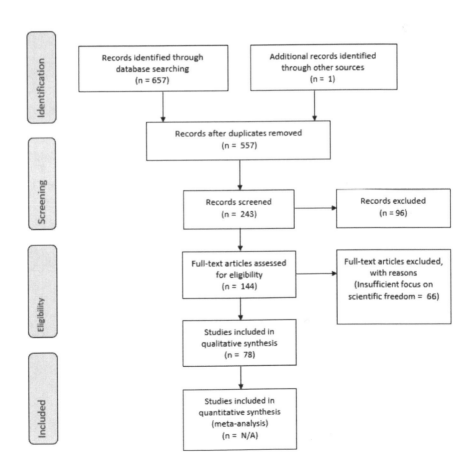

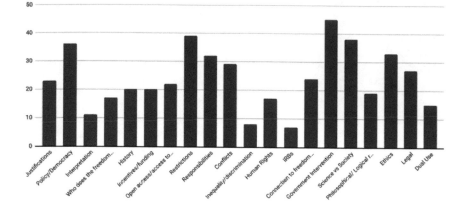

Appendix B: Included Studies

Bayertz, Kurt. 'Three Arguments for Scientific Freedom.' *Ethical Theory and Moral Practice* 9, no. 4 (2006): 377–98.

Benson, Etienne. 'Endangered science: the regulation of research by the US Marine Mammal Protection and Endangered Species Acts.' *Historical Studies in the Natural Sciences* 42, no. 1 (2012): 30–61.

Benson, Sumner. 'Scientific Freedom and National Security: Overcoming Complacency.' Society 23, no. 5 (1986): 12–15.

Bert, Bettina, et al. 'The animal experimentation quandary: Stuck between legislation and scientific freedom: More research and engagement by scientists is needed to help to improve animal welfare without hampering biomedical research.' *EMBO reports* 17, no. 6 (2016): 790–92.

Brown Jr, George E. 'The freedom and the responsibility of investigator-initiated research.' *Academic Medicine* 69, no. 6 (1994): 437–40.

Brown, Mark B., and David H. Guston. 'Science, democracy, and the right to research.' *Science and engineering ethics* 15 (2009): 351–66.

Buchanan, Allen, and Maureen C. Kelley. 'Biodefence and the production of knowledge: rethinking the problem.' *Journal of Medical Ethics* 39, no. 4 (2013): 195–204.

Buchner, Benedikt, et al. 'Tasks, regulations, and functioning of ethics committees.' *Bundesgesundheitsblatt-Gesundheitsforschung-Gesundheitsschutz* 62 (2019): 690–96.

Cantrell, Melissa K. 'International response to Dolly: will scientific freedom get sheared?' *Journal of Law & Health* 13 (1998): 69.

Caulfield, Timothy. 'Scientific freedom and research cloning: can a ban be justified?.' *The Lancet* 364, no. 9429 (2004): 124–26.

Caulfield, Timothy, and Ubaka Ogbogu. 'Stem cell research, scientific freedom and the commodification concern: Vague concerns about possible commodification should not serve as justification to limit the freedom of research.' *EMBO reports* 13, no. 1 (2012): 12–16.

Cave, Jane. 'Political reform and scientific freedom under Gorbachev.' *Technology in Society* 13, no. 1–2 (1991): 69–89.

Chapman, Audrey R. 'Towards an understanding of the right to enjoy the benefits of scientific progress and its applications.' *Journal of Human Rights* 8, no. 1 (2009): 1–36.

Colussi, Ilaria. 'Synthetic biology and freedom os scientific research: a fundamental freedom in front of a new emerging technology.' *Revista de derecho y genoma humano* (2014): 277–87.

Colussi, Ilaria Anna. 'Synthetic biology between challenges and risks: Suggestions for a model of governance and a regulatory framework, based on fundamental rights.' *Review of Law and the Human Genome* 38 (2013): 185–216.

Drenth, Pieter JD. 'Responsible conduct in research.' *Science and Engineering Ethics* 12 (2006): 13–21.

Edsall, John T. 'Scientific Freedom and Responsibility: Report of the AAAS Committee on Scientific Freedom and Responsibility.' *Science* 188, no. 4189 (1975): 687–93.

Efflatoun, H. C. 'Scientific Freedom and Security.' Nature 170, no. 4319 (1952): 215–18.

English, Veronica. 'Autonomy versus protection—who benefits from the regulation of IVF?.' *Human Reproduction* 21, no. 12 (2006): 3044–49.

Esmaeili, Hossein, and Behzad Ataie-Ashtiani. 'The Autonomy of Science as a Civilian Casualty of Economic Warfare: Inadvertent Censorship of Science Resulting from Unilateral Economic Sanctions.' *Science and Engineering Ethics* 27 (2021): 1–9.

Evans, Nicholas G. 'Great expectations—ethics, avian flu and the value of progress.' *Journal of Medical Ethics* 39, no. 4 (2013): 209–13.

Ferguson, James R. 'Scientific inquiry and the first amendment.' *Cornell Law Review* 64 (1978): 639.

Folkers, G. 'On freedom of scientific research.' *Die Pharmazie-An International Journal of Pharmaceutical Sciences* 68, no.7 (2013): 506–20.

Gerhards, Thomas. 'Protest of the Professors. The' Association for Scientific Freedom' in the 1970s.'

Zeitschrift Fur Geschichtswissenschaft 66, no. 9 (2018): 788–90.

Giordano, Simona, John Harris and Lucio Piccirillo. *The freedom of scientific research: Bridging the gap between science and society*. Manchester: Manchester University Press, 2020.

Gostin, Lawrence O. 'Government and science: the unitary executive versus freedom of scientific inquiry.' *Hastings Center Report* 39, no. 2 (2009): 11–12.

Green, Harold P. 'The AEC Proposals—A Threat to Scientific Freedom.' *Bulletin of the Atomic Scientists* 23, no.8 (1967): 15–17.

Green, Harold P. 'The Boundaries of Scientific Freedom.' *Newsletter on Science Technology, & Human Values* 20 (1977).

Guidotti, Tee L. 'Scientific freedom and human rights.' *Archives of Environmental & Occupational Health* 73, no.1 (2018): 1–3.

Harrer, Friedrich. 'Limits of freedom of science' *ALTEX-Alternatives to animal experimentation* 15, no.4 (1998): 199–204.

Horowitz, Irving Louis. 'The politics of physiological psychology Ivan Pavlov's suppressed defense of scientific freedom and its consequences.' *Integrative physiological and behavioral science* 28 (1993): 137–42.

Jonas, Hans. 'Freedom of scientific inquiry and the public interest.' *Hastings Center Report* 6, no. 4 (1976): 15–17.

Knauss, John A. 'Development of the freedom of scientific research issue of the third law of the sea conference.' *Ocean Development & International Law* 1, no.1 (1973): 93–120.

Kress, H. 'Stem cell law of 2002: Sustainable compromise for research on embryonic stem cells?.' *Gynakologe* 36, no.7 (2003): 590–95.

Langenberg, Donald N. 'Scientific freedom and National-Security—Secret knowledge and open inquiry.' *Society* 23, no.5 (1986): 9–12.

Lilienfeld, Scott O. 'When worlds collide: Social science, politics, and the Rind et al. (1998) child sexual abuse meta-analysis.' *American Psychologist* 57, no. 3 (2002): 176–88.

Linden, Belinda. 'Basic Blue Skies Research in the UK: Are we losing out?.' *Journal of Biomedical Discovery and Collaboration* 3 (2008): 1–14.

Little, M. 'Expressing freedom and taking liberties: the paradoxes of aberrant science.' *Medical Humanities* 32, no.1 (2006): 32–37.

Longino, Helen. 'Beyond 'bad science': Skeptical reflections on the value-freedom of scientific inquiry.' *Science, Technology, & Human Values* 8, no.1 (1983): 7–17.

Maechling Jr, Charles. 'Freedom of scientific research: stepchild of the oceans.' *Virginia Journal of International Law* 15, no. 3 (1975): 539–59.

Mainzer, Lewis C. 'Scientific freedom in government-sponsored research.' *The Journal of Politics* 23, no .2 (1961): 212–30.

Marchant, Gary E., and Lynda L. Pope. 'The problems with forbidding science.' *Science and Engineering Ethics* 15 (2009): 375–94.

Marzano, Marco. 'Informed consent, deception, and research freedom in qualitative research.' *Qualitative Inquiry* 13, no.3 (2007): 417–36.

Marzocco, Valeria, and Barbara Salvatore. 'Human dignity and freedom of scientific research: a balancing approach for surplus embryos.' *Biochimica Clinica* 41, no.4 (2017): 368–70.

McCorkle, Richard A. 'Scientific freedom.' *Physics Today* 37, no.3 (1984): 140.

McKelvain, Boyd J. 'Scientific Freedom and National Security: Determining military criticality.' *Society* 23, no.5 (1986): 19–21.

Metzger, Walter P. 'Academic freedom and scientific freedom.' *Daedalus* 107, no. 2(1978): 93–114.

Miller, Seumas, and Michael J. Selgelid. 'Ethical and philosophical consideration of the dual-use dilemma in the biological sciences.' *Science and engineering ethics* 13 (2007): 523–80.

Minker, Jack. 'Computer scientists whose scientific freedom and human rights have been violated: A report of the ACM committee on scientific freedom and human rights.' *Communications of the ACM* 24, no.3 (1981): 134–39.

Minker, Jack, ed. 'Computer professionals whose scientific freedom and human rights have been violated–1982: a report of the ACM committee on scientific freedom and human rights.' *Communications of the ACM* 25, no.12 (1982): 888–94.

Minker, Jack. 'Computer professionals whose scientific freedom and human rights have been violated—1984: a report of the ACM committee on scientific freedom and human rights.' *Communications of the ACM* 28, no.1 (1985): 69–78.

Minker, Jack. 'Scientific freedom and human rights of computer professionals—1989.' *Communications of the ACM* 32, no. 8 (1989): 957–74.

Molnár-Gábor, Fruzsina. 'Germany: a fair balance between scientific freedom and data subjects' rights?.' *Human genetics* 137, no. 8 (2018): 619–26.

Monbiot, George. 'Guard dogs of perception: the corporate takeover of science.' *Science and engineering ethics* 9, no. 1 (2003): 49–57.

Murray, Cherry A., and Saswato R. Das. 'The price of scientific freedom.' *Nature Materials* 2, no. 4 (2003): 204–5.

Ondersma, Steven J. 'Scientific Freedom is Not the Only Issue.' *American Psychologist* 57, no. 2 (2002): 141–42.

Owen, Richard, Phil Macnaghten, and Jack Stilgoe. 'Responsible research and innovation: From science in society to science for society, with society.' *Science and public policy* 39, no.6 (2012): 751–60.

Park, Robert L. 'Restrictions On Scientific Freedom.' *Electro-Culture 1984*. 474. SPIE, 1984.

Pelikan, Jaroslav. 'Can the Arts and Sciences Flourish Even Under Freedom?' *The Polish Review* 40, no.3 (1995): 259–66.

Post, Robert. 'Constitutional Restraints on the Regulations of Scientific Speech and Scientific Research: Commentary on 'Democracy, Individual Rights and the Regulation of Science.'' *Science and engineering ethics* 15 (2009): 431–38.

Probucka, Dorota. 'Ethical Aspects of the Development of Genetic Engineering.' *European Journal of Science and Theology* 14, no. 5 (2018): 69–75.

Pulsinelli, Gary. 'Freedom to explore: Using the Eleventh Amendment to liberate researchers at state universities from liability for intellectual property infringements.' *Washington Law Review* 82 (2007): 275.

Relyea, Harold C. 'Scientific Freedom and Scientific Responsibility.' *Annals of the New York Academy of Sciences* 577, no. 1 (1989): 200–210.

Resnik, David B. 'Openness versus secrecy in scientific research.' *Episteme* 2, no. 3 (2006): 135–47.

Ringeard, Gisele. 'Scientific research: From freedom to deontology.' *Ocean Development & International Law* 1, no. 2 (1973): 121–36.

Ross, David A. 'Threat to the Freedom of Scientific Research in the Deep Sea?.' *Oceanus* 21, no. 1 (1978): 69–71.

Samimian-Darash, Limor, Hadas Henner-Shapira and Tal Daviko. 'Biosecurity as a boundary object: Science, society, and the state.' *Security Dialogue* 47, no. 4 (2016): 329–47.

Sanders, Anselm Kamperman. 'Limits to database protection: fair use and scientific research exemptions.' *Research Policy* 35, no. 6 (2006): 854–74.

Santosuosso, Amedeo, Valentina Sellaroli, and Elisabetta Fabio. 'What constitutional protection for freedom of scientific research?.' *Journal of Medical Ethics* 33, no. 6 (2007): 342–34.

Schachman, Howard K. 'On scientific freedom and responsibility.' *Biophysical chemistry* 100, no. 1–3 (2003): 615–25.

Selya, Rena. 'Defending scientific freedom and democracy: the genetics society of America's response to Lysenko.' *Journal of the History of Biology* 45, no. 3 (2012): 415–42.

Shamoo, Adil E., & David B. Resnik.'Science and Social Responsibility.' In *Responsible Conduct of Research*, 271–92. Oxford: Oxford University Press, 2022.

Siemann, W. 'Prospects and Limits of Scientific Freedom in German Constitutionalism 1815–1918.' *Historisches Jahrbuch* 107, no. 2 (1987): 315–48.

Singer, Peter. 'Ethics and the limits of scientific freedom.' *The Monist* 79, no. 2 (1996): 218–29.

Smith, George P. 'Biotechnology and the Law: Social Responsibility or Freedom of Scientific Inquiry.' *Mercer Law Review* 39 (1987): 437–60.

Thomas, John Meurig. 'Intellectual freedom in academic scientific research under threat.' *Angewandte Chemie International Edition* 22, no. 52 (2013): 5654–55.

Toma, Peter A. 'An Inquiry Into the Scientific Freedom in Soviet Russia Under Khrushchev and Stalin.' *Western Political Quarterly* 14, no. 3 (1961): 50–52.

Victor, Elizabeth. 'Scientific research and human rights: a response to Kitcher on the limitations of inquiry.' *Science and Engineering Ethics* 20, no. 4 (2014): 1045–63.

Wallerstein, Mitchel B. 'Nurturing a dynamic system.' *Society* 23, no. 5 (1986): 22–24.

Weinstein, James. 'Democracy, individual rights and the regulation of science.' *Science and Engineering Ethics* 15 (2009): 407–29.

White, Philip R. 'Grants and Scientific Freedom.' *Science* 132, no. 3440 (1960): 1694–96.

Wilholt, Torsten. 'Scientific freedom: its grounds and their limitations.' *Studies in History and Philosophy of Science Part A* 41, no. 2 (2010): 174–81.

Wozniak, Robert. 'Scientific Freedom and Human Rights of Computer Professionals.' *Communications of the ACM* 32, no. 11 (1989): 1284–85.

Wright, James G., and John H. Wedge. 'Clinicians and patients' welfare: where does academic freedom fit in?.' *BMJ* 329, no. 7469 (2004): 795–96.

Appendix C: List of Examples

Examples	Mentioned in
Italian IVF	Wilholt 2010 Santussuoso 2007
Stem cells/embryos	Wilholt 2010 Singer 1996 Scachman 2003 Santussuoso 2007 Marchant & Pope 2009 Green 1977 Ferguson 1979 Caulfield 2012 Brown 2009
Contraceptives	Wilholt 2010 Marchant & Pope (2009)
Climate change	Wilholt 2010 Maechling 1975
Sex education	Wilholt 2010 Marchant & Pope 2009
Evolution denial	Weinstein 2009 Metzger 1978
Racial genetics/diff	Weinstein 2009 Victor 2014 Singer 1996 Marchant & Pope 2009
Hydrogen bomb	Weinstein 2009 Park 1985 McCorkel 1984
Encryption	Weinstein 2009 Relyea 1989 Park 1985

Examples	Mentioned in
Anti-DRM	Weinstein 2009
	Post 2009
Vulnerable populations	Victor 2014
Tuskegee	Victor 2014
	Singer 1996
	Miller & Selgelid 2008
Animal research	Victor 2014
	Singer 1996
	Bert 2016
	Benson 2012
NIH mandating women in research	Victor 2014
Travel bans	Victor 2014
	Relyea 1989
	Park 1985
	Minker 1981
	Minker 1982
	Minker 1985
	Minker 1989
H5N1	Victor 2014
	Samimian-Darash et al 2016
	Miller & Selgelid 2008
	Evans 2013
Electromagnetism	Thomas 2013
Positron	Thomas 2013
NMR/MRI	Thomas 2013
X-rays	Thomas 2013
Nuclear fission	Thomas 2013
Antibiotics	Thomas 2013
Immunosuppresants	Thomas 2013
Antibodies	Thomas 2013
Molecular structure of DNA	Thomas 2013
Eugenics	Smith 1988
Gene editing	Smith 1988
	Monbiot 2003
	Miller & Selgelid 2008
Beltsville pigs	Smith 1988
Nazi medical trials	Singer 1996
	Santussuoso 2007
	Miller & Selgelid 2008
	Cantrell 1998

Examples	Mentioned in
Milgram/obedience	Singer 1996
Atomic bomb	Singer 1996 Miller & Selgelid 2008 McCorkel 1984 Green 1967
Genetics in crime	Singer 1996
Human genome project	Singer 1996
Central planning of curricula	Siemann 1987
Limitation of movement	Siemann 1987 Selya 2012
Censorship	Siemann 1987 Schachman 2003 Samimian-Darash et al 2016 Miller & Selgelid 2008 Evans 2013
Gervinus treason case	Siemann 1987
Lysenko	Selya 2012 Miller & Selgelid 2008 Marchant & Pope 2009
McCartyism/Loyalty oaths	Selya 2012
JBS Haldane and communism	Selya 2012
US Genetics society / Bellagio resolution	Selya 2012
Post 9/11 US classified research/restrictions	Schachman 2003 Samimian-Darash et al 2016 Resnik 2006 Miller & Selgelig 2008
Cloning	Schachman 2003 Green 1977 Ferguson 1979 Cantrell 1998 Brown 2009
Hertel v Switzerland	Sanders 2006
removing identifiable information from databases	Resnik 2006
Apotex	Resnik 2006
Microfibres (company)	Resnik 2006
Vioxx	Resnik 2006
PNAS milk/botox	Resnik 2006 Marchant & Pope 2009
Smallpox	Miller & Selgelid 2008 Marchant & Pope 2009

Examples	Mentioned in
Synthetic biology	Miller & Selgelid 2008 Marchant & Pope 2009 Colussi 2013 Colussi 2014
Mousepox	Miller & Selgelid 2008 Marchant & Pope 2009
Chemical/biological WMD	Miller & Selgelid 2008 Marchant & Pope 2009 Ferguson 1979
US BioShield	Miller & Selgelid 2008
Project Jefferson (US vaccine-resistant anthrax)	Miller & Selgelid 2008
Cowpox gain of function	Miller & Selgelid 2008
Virulent TB	Miller & Selgelid 2008
Smallpox/Ebola chimera	Miller & Selgelid 2008
Myxoma virus (host gain of function)	Miller & Selgelid 2008
Us Project Clear Vision (biological weapons)	Miller & Selgelid 2008
US biological weapon grenades	Miller & Selgelid 2008
Anthrax genome	Miller & Selgelid 2008
1918 flu/anthrax/other resurrection	Miller & Selgelid 2008 Marchant & Pope 2009
Partisan distribution of funds	Metzger 1978
Selective departmental hiring processes	Metzger 1978
Student pressures	Metzger 1978
Environmental impact statements	Metzger 1978
Human/primate chimeras	Marchant & Pope 2009
"Only give enough freedom as needed"	Mainzer 1961
Bending rules	Longino 1983
Data fudging	Longino 1983
Values in science	Longino 1983
Scientific Misconduct	Longino 1983 Little 2006 Jonas 1976 Frankl 1994 Ferguson 1979 Evans 2013 Drenth 2006
Notion of freedom and taking liberties	Little 2006
Blue Skies research	Linden 2008

Examples	Mentioned in
Coastal states against research	Maechling 1975 Knauss 1973
Pollution	Maechling 1975 Knauss 1973
Offshore resources	Maechling 1975 Knauss 1973
Overfishing	Maechling 1975 Knauss 1973
Abuse of power	Gostin 2009
Democrats v Republicans	Gostin 2009 Brown 2009
Accountability	Frankl 1994
Boundaries/ limits of Scientific freedom	Frankl 1994 Caulfield 2004
Dental Research	Frankl 1994
Education	Folkers 2013
Bio-terrorism	Ferguson 1979
Censorship to avoid misuse	Evans 2013
IVF	English 2006
Nuclear advent	Efflatoun 1952
Anglo-american Relationship	Efflatoun 1952
Information sources	Esdall 1975
Value free v Value Bound	Drenth 2006
Fraud	Drenth 2006
Narrowing and broadening of academic freedom	Colussi 2014
Russia	Cave 1991
Moral status of an embyro	Caulfield 2004
Single Ideological View	Caulfield 2004
Using fear to write policy agains science	Caulfield 2012
When inquiry is encumbered	Cantrell 1998
When Science goes awry	Cantrell 1998
Responsibilities of ethics committiees	Buchner 2019
Biodefense	Buchanan 2013
Dual-use opportunity	Buchanan 2013
Role of institutions/institutional design	Buchanan 2013
Social epistemology Politicization of science	Buchanan 2013 Brown 2009

Examples	Mentioned in
Freedom of Science=Academic freedom<expressive conduct	Brown 2009
Gap between rich and poor	Brown 1994
Basic v applied research	Brown 1994
Capitalism v utilitarianism	Brown 1994
Aim of research	Brown 1994
Population growth	Brown 1994
Gene modification in animals	Bert 2016
International Collaboration	Bert 2016
Political misalignment	Benson 1986
American Propaganda	Benson 1986
Scientific application	Benson 1986
Militaristic application	Benson 1986
USA V USSR/Russia	Benson 1986
Arms race	Benson 1986
Animal Welfare	Bert 2016 Benson 2012
Cognition V Action	Benson 2012 Bayertz 2016
Wildlife protection & activism	Benson 2012
Protected species	Benson 2012
Aristotelian Argument	Bayertz 2016
Kantian Argument	Bayertz 2016
Baconian Argument	Bayertz 2016

Notes

ARTIFICIAL INTELLIGENCE / INTRODUCTION

1. The human right to 'enjoy the benefits of the progress of science and its applications' is protected by Article 15(1)b of the International Covenant on Economic, Social and Cultural Rights; Article 15(3) of the same Covenant protects the "freedom indispensable for scientific research and creative activity." The right to science also appears in the Universal Declaration of Human Rights (Article 27(1)) and the American Declaration of the Rights and Duties of Man (Article XIII). Much more will be said about this right later in this chapter and in Chapter Three of Part I of this book.

2. Sebastian Porsdam Mann, Helle Porsdam, and Yvonne Donders, "'Sleeping Beauty': The Right to Science as a Global Ethical Discourse," *Human Rights Quarterly* 42, no. 2 (2020): 332–56.

3. Jürgen Schmidhuber, "Deep learning in neural networks: An overview." *Neural Networks* 61 (2015): 85-117.

4. Douglas Kell, "Mission-Driven Research Is No Substitute for Discovery Science," Times Higher Education, May 7, 2020, https://www.timeshighereducation.com/opinion/mission-driven-research-no-substitute-discovery-science.

5. Kell. In an interview on CBS Morning, Hinton explains that his and Yoshua Bengio's research was funded by basic, curiosity-driven research funds by the Canadian government, including through the Canadian Institute for Advanced Research, https://www.youtube.com/watch?v=qpoRO378qRY.

6. Ben Wodecki, "ChatGPT Passes 1 Billion Page Views," AI Business, March 10, 2023, https://aibusiness.com/nlp/chatgpt-passes-1b-page-views.

7. Technically, LLMs predict tokens rather than words. Tokens are a smaller unit of language which can include full words, but can also be fragments of words, or punctuation marks. Shanahan describes how LLMs function as follows: 'Here's a fragment of text. Tell me how this fragment might go on. According to your model of the statistics of human language, what words are likely to come next?' Murray Shanahan, 'Talking About Large Language Models,' arXiv, 16 February 2023, https://doi.org/10.48550/arXiv.2212.03551.

8. Lakshmi Varanasi, "List: Here Are the Exams ChatGPT Has Passed so Far," March 21, 2023, https://www.businessinsider.com/list-here-are-the-exams-chatgpt-has-passed-so-far-2023-1?r=US&IR=T.

9. Ben Dickson, "What to Know about the Applications of GPT-4 - TechTalks," March 20, 2023, https://bdtechtalks.com/2023/03/20/gpt-4-applications-limits/.

10. Luke Larsen, "5 Amazing Things GPT-4 Has Already Done That Show Its Power," Digital Trends, March 30, 2023, https://www.digitaltrends.com/computing/gpt-4-amazing-things-show-its-power/.

11. Emily Dreibelbis, "Everyone Is Writing a Novel, Even ChatGPT," PCMag UK, February 21, 2023, https://uk.pcmag.com/news/145544/everyone-is-writing-a-novel-even-chatgpt.

12. Bommasini et al., "On the Opportunities and Risks of Foundation Models," arXiv, August 16, 2021, https://arxiv.org/abs/2108.07258.

13. The prompt used to generate the image was: "Scientific Freedom: Heart of the Right to Science." The original result contained different symbols in the oval images in each of the corners of the image; we replaced these with a symbol of science chosen by each of the authors and also generated with Midjourney. The prompts used to generate these symbols were (clockwise, starting from top left): "an atom, scientific intricate illustration, subdued colors, painting --v 5 " "dna helix painting, subdued colors, --s 500 --c 3"; "The open book of science and knowledge, owl, blue sky, calm color palette, calm, digital art --s 1000 --c 30"; "Science, Research, Freedom, calm color palette, cover art, --s 750 --c 9."

14. Sebastian Porsdam Mann et al., "Generative AI entails a credit–blame asymmetry" *Nature Machine Intelligence* 5 (2023): 472–75, https://doi.org/10.1038/s42256-023-00653-1.

15. Shanahan, "Talking About Large Language Models."

16. Stefan Harrer, "Attention is not all you need: the complicated case of ethically using large language models in healthcare and medicine," *eBioMedicine* 90 (2023): 104512.

17. Emily M. Bender et al., "On the Dangers of Stochastic Parrots: Can Language Models Be Too Big?" *FAccT '21: Proceedings of the 2021 ACM Conference on Fairness, Accountability, and Transparency* (March 2021): 610–23.

18. Ali Borji, "A categorical archive of ChatGPT failures" arXiv, April 3, 2023, https://arxiv.org/abs/2302.03494.

19. John Jumper et al., "Highly accurate protein structure prediction with Alpha-Fold," *Nature* 596 (2021): 583–89.

20. Yangyang Chen et al., "Deep generative model for drug design from protein target sequence." *Journal of Cheminformatics* 15, no. 1 (2023): 38.

21. Fabio Urbina et al., "Dual use of artificial-intelligence-powered drug discovery," *Nature Machine Intelligence* 4, no. 3 (2022): 189–91.

22. The open letter was published by the Future of Life institute, https://futureoflife.org/open-letter/pause-giant-ai-experiments/

23. American Association for the Advancement of Science, Statement on "Scientific Freedom and Responsibility," 2017, accessed April 9, 2023, https://www.aaas.org/programs/scientific-responsibility-human-rights-law/aaas-statement

-scientific-freedom#:~:text=Scientific%20freedom%20is%20the%20freedom,in %20accordance%20with%20scientific%20responsibility.

24. John M. Thomas, "Intellectual Freedom in Academic Scientific Research under Threat," *Angewandte Chemie-International Edition* 52, no. 22 (2013): 5654–55, https://doi.org/10.1002/anie.201302192; Mariana Mazzucato, *Mission Economy* (London: Allen Lane, 2021); Donald W. Braben, *Scientific Freedom: The Elixir of Civilization* (New Jersey: John Wiley & Sons, 2007).

25. The Golden Fleece Award (1975–1988) was an award given to public officials in the United States by U.S. Senator Edward William Proxmire for what he considered to be wasteful use of public money.

26. Lucie Laplane et al., "Why Science Needs Philosophy," *Proceedings of the National Academy of Sciences* 116, no. 10 (5 March 2019): 3948–52, https://doi .org/10.1073/pnas.1900357116; Regina Moirano, Marisa Analía Sánchez, and Libor Štěpánek, "Creative Interdisciplinary Collaboration: A Systematic Literature Review," *Thinking Skills and Creativity* 35 (1 March 2020): 100626, https://doi.org/10.1016/j .tsc.2019.100626; Jian Wang, Bart Thijs, and Wolfgang Glänzel, "Interdisciplinarity and Impact: Distinct Effects of Variety, Balance, and Disparity," *PLoS ONE* 10, no. 5 (22 May 2015): e0127298, https://doi.org/10.1371/journal.pone.0127298.

27. Johan S. G. Chu and James A. Evans, "Slowed Canonical Progress in Large Fields of Science," *Proceedings of the National Academy of Sciences* 118, no. 41 (12 October 2021): e2021636118, https://doi.org/10.1073/pnas.2021636118; Michael Park, Erin Leahey, and Russell J. Funk, "Papers and Patents Are Becoming Less Disruptive over Time," *Nature* 613, no. 7942 (January 2023): 138–44, https://doi.org/10 .1038/s41586-022-05543-x.

28. Klaus D. Beiter, "Where Have All the Scientific and Academic Freedoms Gone? And What Is 'Adequate for Science'? The Right to Enjoy the Benefits of Scientific Progress and Its Applications," *Israel Law Review* 52, no. 2 (July 2019): 233–91; Donald W. Braben, *Scientific Freedom: The Elixir of Civilization* (San Francisco: Stripe Press, 2020).

29. United Nations Human Rights Office of the High Commissioner, "Instruments and mechanisms: International Human Rights Law," accessed March 16, 2023, https: //www.ohchr.org/en/instruments-and-mechanisms/international-human-rights-law.

30. Sebastian Porsdam Mann, Yvonne Donders, and Helle Porsdam, "The Right to Science in Practice: A Step in Four Stages," in *The Right to Science: Then and Now* eds. Helle Porsdam and Sebastian Porsdam Mann (Cambridge: Cambridge University Press, 2022), 231-45.

CHAPTER 1

1. Gundling, 1722, quoted in Torsten Wilholt, 'Scientific freedom,' *Studies in History and Philosophy of Science* 41, no. 2 (June 2010): 175.

2. Wolf, quoted in Torsten Wilholt, 'Scientific freedom,' *Studies in History and Philosophy of Science* 41, no. 2 (June 2010): 175.

3. Isaac Newton famously wrote in a 1675 letter to fellow scientist Robert Hooke that, 'if I have seen further, it is by standing on the shoulders of giants.' See, e.g., Chaomei Chen 'On the Shoulders of Giants,' In *Mapping Scientific Frontiers: The Quest for Knowledge Visualization* (London: Springer, 2003), 135–66.

4. See Walter P. Metzger, 'Academic Freedom and Scientific Freedom' *Daedalus* 107, no. 2 (Spring 1978): 93–114.

5. Denise Phillips, 'Francis Bacon and the Germans: Stories from when 'science' meant 'Wissenschaft.'' *History of Science* 53, no. 4 (2015), 378–94.

6. See Metzger, 'Academic freedom and scientific freedom.'

7. Metzger.

8. Metzger.

9. Walter P. Metzger, 'The 1940 Statement of Principles on Academic Freedom and Tenure.' *Law and Contemporary Problems* 53, no. 3 (1990): 3–77, https://doi.org/10.2307/1191793; David Randall, 'Charting Academic Freedom,' *National Association of Scholars* 2018, https://www.nas.org/reports/charting-academic-freedom-103-years-of-debate.

10. United Nations, 'Vienna Convention on the Law of Treaties' (Adopted 23 May 1969, Entry into Force 27 January 1980) 1155 UNTS 331 (VCLT), 1969, Art. 18: 'A State is obliged to refrain from acts which would defeat the object and purpose of a treaty when: (a) it has signed the treaty or has exchanged instruments constituting the treaty subject to ratification, acceptance or approval, until it shall have made its intention clear not to become a party to the treaty; or (b) it has expressed its consent to be bound by the treaty, pending the entry into force of the treaty and provided that such entry into force is not unduly delayed.'

11. Amedeo Santosuosso, Valentina Sellaroli and Elisabetta Fabio, 'What Constitutional Protection for Freedom of Scientific Research?' *Journal of Medical Ethics* 33, no. 6 (June 2007): 342–44, https://doi.org/10.1136/jme.2007.020594; Cesare P. R. Romano and Andrea Boggio, 'Right to Science,' in *Max Planck Encyclopedia of Comparative Constitutional Law* (October 2020); Klaus D. Beiter, Terence Karran, and Kwadwo Appiagyei-Atua, ''Measuring' the Erosion of Academic Freedom as an International Human Right: A Report on the Legal Protection of Academic Freedom in Europe.' *Vanderbilt Journal of Transnational Law* 49, no. 3 (2016): 597–691, https://papers.ssrn.com/abstract=3531682.

12. J. Weinstein, 'Democracy, Individual Rights and the Regulation of Science.' *Science and Engineering Ethics* 15, no. 3 (2009): 407–29. https://doi.org/10.1007/s11948-009-9145-2.

13. *Keyishian v Board of Regents* (385 U.S. 589 1967).

14. *Keyishian v Board of Regents*, see also: Robert Post, 'Constitutional Restraints on the Regulations of Scientific Speech and Scientific Research.' *Science and Engineering Ethics* 15, no. 3 (2009): 431–38.

15. 'Science,' Merriam-Webster.com Dictionary, accessed May 5, 2023. https://www.merriam-webster.com/dictionary/science.

16. See Philipps, note 5 above.

17. Entscheidungen des Bundesverfassungsgerichts, Band 35, S. 113; authors' own translation (jede Tätigkeit, die „nach Inhalt und Form als ernsthafter planmäßiger Versuch zur Ermittlung der Wahrheit anzusehen ist').

18. Philipp Roelli, *Latin as the Language of Science and Learning.* (Berlin: De Gruyter, 2021), 21; Samer Akkach, ed., *Ilm: Science, Religion and Art in Islam* (Adelaide: University of Adelaide Press, 2019); Yiming Zhang and Zengyi Zhang "Kexue Wenhua' in Chinese and 'Scientific Culture,' 'Science Culture,' 'Culture of Science' and 'Science as Culture' in English: The Meanings and the Structure.' *Cultures of Science* 1, no. 1 (2018), 25.

19. See Wilholt, note 1 above, at 175.

20. Wilholt, 175.

21. Søren Holm, 'Religion and Scientific Freedom.' In *Scientific Freedom: An Anthology on Freedom of Scientific Research*, edited by S. Giordano, J. Coggon and M. Cappato, 129. London: Bloomsbury Academic, 2012.

22. Article 4 ICESCR: 'The States Parties to the present Covenant recognize that, in the enjoyment of those rights provided by the State in conformity with the present Covenant, the State may subject such rights only to such limitations as are determined by law only in so far as this may be compatible with the nature of these rights and solely for the purpose of promoting the general welfare in a democratic society.' The meaning and relevance of these criteria are explored in greater depth in the following chapter and in Part II, Chapter One.

23. See Klaus D. Beiter, 'Where Have All the Scientific and Academic Freedoms Gone? And What Is 'Adequate for Science'? The Right to Enjoy the Benefits of Scientific Progress and Its Applications.' *Israel Law Review* 52, no. 2 (July 2019): 244, quoting Eric Barendt, *Academic Freedom and the Law—A Comparative Study* (London: Bloomsbury Publishing, 2010), 37.

24. Beiter, 248.

CHAPTER 2

1. William W. Van Alstyne, 'Academic Freedom and the First Amendment in the Supreme Court of the United States: An Unhurried Historical Review,' *Law and Contemporary Problems* 53, no. 3 (1990): 79.

2. Walter P. Metzger, 'Academic Freedom and Scientific Freedom,' *Daedalus* 107, no. 2 (Spring 1978): 93–114.

3. Klaus D. Beiter, 'Where Have All the Scientific and Academic Freedoms Gone? And What Is 'Adequate for Science'? The Right to Enjoy the Benefits of Scientific Progress and Its Applications.' *Israel Law Review* 52, no. 2 (July 2019): 233–91.

4. John Milton, 'Areopagitica' (1644) Available: https://milton.host.dartmouth.edu /reading_room/areopagitica/text.html

5. Beiter, 'Where Have All the Scientific and Academic Freedoms Gone?' 238.

6. Belimda Linden, 'Basic Blue Skies Research in the UK: Are We Losing Out?' *Journal of Biomed Discovery and Collab* 3 (29 February 2008): 3, https://doi.org /10.1186/1747-5333-3-3; John Meurig Thomas, Intellectual Freedom in Academic

Scientific Research under Threat.' *Angewandte Chemie International Edition* 22, no. 52 (2013): 5654–55.

7. Torsten Wilholt, 'Scientific Freedom: Its Grounds and Their Limitations,' *Studies in History and Philosophy of Science* 41, no. 2 (June 2010): 174 https://doi.org/10.1016/j.shpsa.2010.03.003.

8. Wilholt, 174.

9. Wilholt,

10. Zachary Pirtle and Jared Moore, 'Where Does Innovation Come From?: Project Hindsight, TRACEs, and What Structured Case Studies Can Say About Innovation' *IEEE Technology and Society Magazine* 38, no. 3 (September 2019): 56–67, https://doi.org/10.1109/MTS.2019.2930270.

11. Jonathan Haskel and Gavin Wallis, 'Public Support for Innovation, Intangible Investment and Productivity Growth in the UK Market Sector,' *IZA Discussion Papers* (February 2010), https://ideas.repec.org//p/iza/izadps/dp4772.html.

12. Haskel and Wallis.

13. Jamie Shaw, 'On the Very Idea of Pursuitworthiness,' *Studies in History and Philosophy of Science* 91 (1 February 2022): 103–12. https://doi.org/10.1016/j.shpsa.2021.11.016.

14. Jamie Shaw, 'There and Back Again: Revisiting Vannevar Bush, the Linear Model, and the Freedom of Science,' *Research Policy* 51, no. 10 (1 December 2022): 104610. https://doi.org/10.1016/j.respol.2022.104610.

15. Shaw; Mariana Mazzucato, *Mission Economy* (London: Allen Lane, 2021).

16. We shall come back to the issue of funding and its impact on scientific freedom in Chapter Four.

17. See e.g., Eric Barendt, *Academic Freedom and the Law—A Comparative Study* (London: Bloomsbury Publishing, 2010); Wilholt, 'Scientific Freedom,'

18. *Sweezy v. New Hampshire* 354 US 234 (1957)

19. Klaus D. Beiter, Terence Karran, and Kwadwo Appiagyei-Atua, 'Yearning to Belong: Finding a Home for the Right to Academic Freedom in the U.N. Human Rights Covenants,' *Intercultural Human Rights Law Review* 11 (2016): 107–90.

20. UN Committee on Economic, Social and Cultural Rights, General Comment No 13: The Right to Education, UN Doc E/C.12/1999/10 (1999), para. 38

21. Barendt, *Academic Freedom and the Law*; Wilholt, 'Scientific Freedom,'

22. Amedeo Santosuosso, Valentina Sellaroli, and Elisabetta Fabio, 'What Constitutional Protection for Freedom of Scientific Research?,' *Journal of Medical Ethics* 33, no. 6 (June 2007): 342–44; Wilholt, 'Scientific Freedom,'

23. Mark B. Brown and David H. Guston, 'Science, Democracy, and the Right to Research,' *Science and Engineering Ethics* 15, no. 3 (September 2009): 351–66, https://doi.org/10.1007/s11948-009-9135-4; Timothy Caulfield and Ubaka Ogbogu, 'Stem Cell Research, Scientific Freedom and the Commodification Concern,' *Embo Reports* 13, no. 1 (January 2012): 12–16, https://doi.org/10.1038/embor.2011.232; James R. Ferguson, 'Scientific Inquiry and the First Amendment,' *Cornell Law Rev* 64, no. 4 (April 1979): 639–65.; Harold P. Green, 'III. the Boundaries of Scientific Freedom,' *Newsletter on Science, Technology & Human Values* 2, no. 3 (1 June 1977): 17–21, https://doi.org/10.1177/016224397700200303; Gary E. Marchant and

Lynda L. Pope, 'The Problems with Forbidding Science,' *Science and Engineering Ethics* 15, no. 3 (September 2009): 375–94, https://doi.org/10.1007/s11948-009-9130 -9; Peter Singer, 'Ethics and the Limits of Scientific Freedom,' *The Monist* 79, no. 2 (1996).

24. Charles Jr. Maechling 'Freedom of Scientific Research; Stepchild of the Oceans,' *Virginia Journal of International Law* 5 (1975): 539–59; Metzger, 'Academic Freedom and Scientific Freedom'; Wilholt, 'Scientific Freedom,'

25. Marchant and Pope, 'The Problems with Forbidding Science'; Wilholt, 'Scientific Freedom.'

26. Marchant and Pope; Singer, 'Ethics and the Limits of Scientific Freedom'; Elizabeth Victor, 'Scientific Research and Human Rights: A Response to Kitcher on the Limitations of Inquiry.' *Science and Engineering Ethics* 20, no. 4 (December 2014): 1045–63, https://doi.org/10.1007/s11948-013-9497-5; J. Weinstein, 'Democracy, Individual Rights and the Regulation of Science,' *Science and Engineering Ethics* 15, no. 3 (2009): 407–29. https://doi.org/10.1007/s11948-009-9145-2.

27. Jack Minker, 'Computer Scientists Whose Scientific Freedom and Human Rights Have Been Violated: A Report of the ACM Committee on Scientific Freedom and Human Rights,' *Communications of the Acm* 24, no. 3 (1981): 134–39, https://doi.org/10.1145/358568.358584; Robert L. Park, 'Restrictions on Scientific Freedom,' *Proceedings of the Society of Photo-Optical Instrumentation Engineers Electro-Culture* 474 (1984): 64–67. https://doi.org/10.1117/12.942475; Harold C. Relyea, 'Scientific Freedom and Scientific Responsibility,' *Annals of the New York Academy of Sciences* 577, no. 1 (December 1989): 200–210, https://doi.org/10 .1111/j.1749-6632.1989.tb15066.x; Victor, 'Scientific Research and Human Rights,'

28. Santosuosso, Sellaroli, and Fabio, 'What Constitutional Protection for Freedom of Scientific Research?'; George P. Smith, 'Biotechnology and the Law: Social Responsibility or Freedom of Scientific Inquiry?' *Mercer Law Rev* 39, no. 2 (Winter 1988): 437–60.

29. Seumas Miller and Michael J. Selgelid, 'Ethical and Philosophical Consideration of the Dual-Use Dilemma in the Biological Sciences,' *Science and Engineering Ethics* 13, no. 4 (December 2007): 523–80, https://doi.org/10.1007/s11948-007-9043 -4; George Monbiot, 'Guard Dogs of Perception: The Corporate Takeover of Science,' *Science and Engineering Ethics* 9, no. 1 (January 2003): 49–57, https://doi.org/10 .1007/s11948-003-0019-8; Smith, 'Biotechnology and the Law.'

30. Miller and Selgelid, 'Ethical and Philosophical Consideration,'; Limor Samimian-Darash, Hadas Henner-Shapira, and Tai Daviko, 'Biosecurity as a Boundary Object: Science, Society, and the State,' *Security Dialogue* 47, no. 4 (August 2016): 329–47, https://doi.org/10.1177/0967010616642918; Howard K. Schachman, 'On Scientific Freedom and Responsibility,' *Biophysical Chemistry* 100, no. 1–3 (2003): 615–25, https://doi.org/10.1016/s0301-4622(02)00310-1; Rena Selya, 'Defending Scientific Freedom and Democracy: The Genetics Society of America's Response to Lysenko,' *Journal of the History of Biology* 45, no. 3 (August 2012): 415–42, https://doi.org/10.1007/s10739-011-9288-2.

31. Marchant and Pope, 'The Problems with Forbidding Science'; Miller and Selgelid, 'Ethical and Philosophical Consideration'; Selya, 'Defending Scientific Freedom and Democracy.'

32. Selya.

33. Brown and Guston, 'Science, Democracy, and the Right to Research'; Melissa K. Cantrell, 'International Response to Dolly: Will Scientific Freedom Get Sheared?,' *Journal of Law Health* 13, no. 1 (1998): 69–102; Ferguson, 'Scientific Inquiry,'; Green, 'III. the Boundaries of Scientific Freedom'; Schachman, 'On Scientific Freedom and Responsibility.'

34. Ilaria Anna Colussi, 'Synthetic Biology and the Freedom of Scientific Research: A Fundamental Freedom in Front of a New Emerging Technology.' *Rev Derecho Genoma Hum* Spec No (2014): 277–87; Colussi, 'Synthetic Biology between Challenges and Risks: Suggestions for a Model of Governance and a Regulatory Framework, Based on Fundamental Rights,' *Review of Law and the Human Genome*, no. 38 (June 2013): 185–214; Marchant and Pope, 'The Problems with Forbidding Science'; Miller and Selgelid, 'Ethical and Philosophical Consideration.'

35. Metzger, 'Academic Freedom and Scientific Freedom.'

36. Metzger.

37. Barendt, *Academic Freedom and the Law*; Beiter, 'Where Have All the Scientific and Academic Freedoms Gone?'; Marco Marzano, 'Informed Consent, Deception, and Research Freedom in Qualitative Research—A Cross-Cultural Comparison,' *Qualitative Inquiry* 13, no. 3 (April 2007): 417–36, https://doi.org/10.1177/1077800406297665; Boyd J. McKelvain, 'Scientific Freedom and National Security—Determining Military Criticality.' *Society* 23, no. 5 (August 1986): 19–21, https://doi.org/10.1007/bf02695552; Miller and Selgelid, 'Ethical and Philosophical Consideration,'; Richard Owen, Phil Macnaghten, and Jack Stilgoe, 'Responsible Research and Innovation: From Science in Society to Science for Society, with Society.' *Science and Public Policy* 39, no. 6 (2012): 751–60. https://doi.org/10.1007/s11948-016-9782-1; Singer, 'Ethics and the Limits of Scientific Freedom.'

38. Christian Starck, 'Freedom of Scientific Research and Its Restrictions in German Constitutional Law.' *Israel Law Review* 39, no. 2 (ed 2006): 114, https://doi.org/10.1017/S0021223700013030.

39. Quoted in Starck p. 115.

40. Quoted in Starck p. 115.

41. Quoted in Starck p.115.

42. See e.g., Barendt, *Academic Freedom and the Law*.

43. Sebastian Porsdam Mann et al. 'The Human Right to Enjoy the Benefits of the Progress of Science and Its Applications.' *The American Journal of Bioethics* 17, no. 10 (3 October 2017): 34–36, https://doi.org/10.1080/15265161.2017.1365194.

44. Mazzucato, *Mission Economy*.

45. Harold P. Green, 'The AEC Proposals—A Threat To Scientific Freedom.' *Bulletin of the Atomic Scientists* 23, no. 8 (1967): 15–17. https://doi.org/10.1080/00963402.1967.11455114; Richard A. McCorkle, 'Scientific Freedom' *Physics Today* 37, no. 3 (1984): 140–42, https://doi.org/10.1063/1.2916146; Park, 'Restrictions on Scientific Freedom'; Singer, 'Ethics and the Limits of Scientific Freedom'; Thomas,

'Intellectual Freedom in Academic Scientific Research,'; Weinstein, 'Democracy, Individual Rights and the Regulation of Science.'

46. Nicholas G. Evans, 'Great Expectations-Ethics, Avian Flu and the Value of Progress' *Journal of Medical Ethics* 39, no. 4 (April 2013): 209–13. https://doi.org /10.1136/medethics-2012-100712; Miller and Selgelid, 'Ethical and Philosophical Consideration,'; Samimian-Darash, Henner-Shapira, and Daviko, 'Biosecurity as a Boundary Object,'; Victor, 'Scientific Research and Human Rights,.'

47. Miller and Selgelid; David B. Resnik, 'Openness versus Secrecy in Scientific Research Abstract,' *Episteme* 2, no. 3 (1 February 2006): 135–47; Samimian-Darash, Henner-Shapira, and Daviko; Schachman, 'On Scientific Freedom and Responsibility.'

48. Smith, 'Biotechnology and the Law,'; Monbiot, 'Guard Dogs of Perception,'; Miller and Selgelid; Marchant and Pope, 'The Problems with Forbidding Science'; Colussi, 'Synthetic Biology and the Freedom of Scientific Research,'; Colussi, 'Synthetic Biology between Challenges and Risks,.'

49. Miller and Selgelid; Marchant and Pope.

50. Samimian-Darash, Henner-Shapira, and Daviko, 'Biosecurity as a Boundary Object,' 334.

51. Lawrence O. Gostin, 'Government and Science: The Unitary Executive versus Freedom of Scientific Inquiry.' *Hastings Center Report* 39, no. 2 (April 2009): 11–12, https://doi.org/10.1353/hcr.0.0114.; Schachman, 'On Scientific Freedom and Responsibility'; Samimian-Darash, Henner-Shapira, and Daviko,; Resnik, 'Openness versus Secrecy,'; Miller and Selgelid, 'Ethical and Philosophical Consideration,.'

52. Beatrice M. Dias et al., 'Effects of the USA PATRIOT Act and the 2002 Bioterrorism Preparedness Act on Select Agent Research in the United States' *Proceedings of the National Academy of Sciences of the United States of America* 107, no. 21 (25 May 2010): 9556–61, https://doi.org/10.1073/pnas.0915002107.

53. e.g., Marchant and Pope, 'The Problems with Forbidding Science'; John T. Edsall, 'Scientific Freedom and Responsibility,' *Science* 188, no. 4189 (16 May 1975): 687–93, https://doi.org/10.1126/science.11643270.

54. Marchant and Pope, 'The Problems with Forbidding Science.'

55. Quoted in Marchant and Pope p. 382.

56. Quoted in Marchant and Pope p. 382.

57. Quoted in Marchant and Pope p. 384.

58. Marchant and Pope, 385.

59. Marchant and Pope.

60. Marchant and Pope, 381.

61. Marchant and Pope.

62. Marchant and Pope, 386.

63. See Dave Levitan's *Not a Scientist*. (New York: W. W. Norton & Company, 2017) for more on this.

64. 'Indispensable, adj. and n.' Oxford English Dictionary, accessed may 12, 2023. Oxford University Press,https://www.oed.com/viewdictionaryentry/Entry /94574?p=emailAO8N93U1M0kvk&d=94574; 'Indispensable' Cambridge Dictionary Online, accessed may 12, 223. Cambridge University Press: https://dictionary .cambridge.org/dictionary/english/indispensable.

CHAPTER 3

1. Bishop, 'Foreword', xi.

2. Lea Bishop, 'Foreword', in *The Right to Science: Then and Now* eds. Helle Porsdam & Sebastian Porsdam Mann (Cambridge: Cambridge University Press, 2022), xi-xii.

3. Mikel Mancisidor, 'The Dawning of a Right,' in *The Right to Science: Then and Now* eds. Helle Porsdam & Sebastian Porsdam Mann (Cambridge: Cambridge University Press, 2022), 18.

4. Mancisidor, 18.

5. Herbert Hoover, 'Address at Madison Square Garden in New York City,' October 31, 1932, https://www.presidency.ucsb.edu/documents/address-madison-square -garden-new-york-city-2

6. Joseph P. Harris, 'Research: A National Resource. Report of the Science Committee to the National Resources Committee,' *American Political Science Review* 33, no. 3 (June 1939): 506–7, https://doi.org/10.2307/1948809.

7. A. Hunter Dupree, *Science in the Federal Government: A History of Policies and Activities to 1940* (Cambridge: Harvard University Press, 1957) pp. 350–58.

8. Elizabeth Borgwardt, *A New Deal for the World: America's Vision for Human Rights* (Cambridge: Harvard University Press, 2005).

9. Leo E. Strine, 'Made for This Moment: The Enduring Relevance of Adolf Berle's Belief in a Global New Deal,' *Seattle University Law Review* 42, no. 2 (2019).

10. Strine.

11. The various drafts are accessible online via the FDR Library: http:// www.fdrlibrary.marist.edu/archives/collections/franklin/index.php?p=collections/ findingaid&id=582. The second draft is found in speech file 1353A, box 58, and can be accessed here: http://www.fdrlibrary.marist.edu/_resources/images/msf/msf01407.

12. The fifth draft is found in speech file 1353C, box 58, and can be accessed here: http://www.fdrlibrary.marist.edu/_resources/images/msf/msf01409.

13. Mary Ann Glendon, 'The Forgotten Crucible: The Latin American Influence on the Universal Human Rights Idea,' *Harvard Human Rights Journal* 16 (2003): 28.

14. Glendon, 28.

15. Cesare P. R. Romano, 'The Origins of the Right to Science: The American Declaration on the Rights and Duties of Man,' in *The Right to Science: Then and Now*, eds. Helle Porsdam and Sebastian Porsdam Mann (Cambridge: Cambridge University Press, 2021), 35, https://doi.org/10.1017/9781108776301.004.

16. Paolo Carozza, 'From Conquest to Constitutions: Retrieving a Latin American Tradition of the Idea of Human Rights,' *Human Rights Quarterly* 25 (1 January 2003): 281–313; Glendon, 'The Forgotten Crucible.'

17. Inter-American Juridical Committee, 'Draft Declaration of the International Rights and Duties of Man and Accompanying Report, Formulated by the Inter-American Juridical Committee in Accordance with Resolutions IX and XL of the Inter-American Conference on Problems of War and Peace Held at Mexico City, February 21–March 8, 1945' (Pan American Union, 1946).

18. Inter-American Juridical Committee.

19. Inter-American Juridical Committee.

20. Inter-American Juridical Committee.

21. Inter-American Juridical Committee, 48.

22. Fifteenth meeting of the First Session of the Drafting Committee, June 23, 1947; Ninth meeting of the Working Group on the Draft Declaration during the Second Session of the Commission of Human Rights, December 10, 1947; 70th meeting of the Commission of Human Rights during its Second Session, June 11, 1948; 150th–152nd meetings of the Third Committee of the UN General Assembly at its Third Session, November 20th and 22nd, 1948.

23. 228th meeting of the Commission on Human Rights, at its Seventh Session, June 28, 1951; 292nd–294th meetings of the Commission on Human Rights, at its Eighth Session, May 27, 1952; 795th–799th meetings of the Twelfth Session of the General Assembly, October 30–November 1 and November 4, 1957.

24. Sebastian Porsdam Mann, 'The Right to Science or to Wissenschaft? A Chronology and Five Lessons from the Travaux Préparatoires,' *SSRN* (November 2022), https://papers.ssrn.com/sol3/papers.cfm?abstract_id=4282820; Tara Smith, 'Understanding the Nature and Scope of the Right to Science through the Travaux Préparatoires of the Universal Declaration of Human Rights and the International Covenant on Economic, Social and Cultural Rights,' *The International Journal of Human Rights* 24, no. 8 (2020): 1156–79; William Schabas, 'Looking Back: How the Founders Considered Science and Progress in Their Relation to Human Rights,' *European Journal of Human Rights* 2015, no. 4 (2015); Aurora Plomer, 'IP Rights and Human Rights: What History Tells Us and Why It Matters,' in *The Right to Science: Then and Now*, ed. Helle Porsdam and Sebastian Porsdam Mann (Cambridge: Cambridge University Press, 2021), 54–75, https://doi.org/10.1017/9781108776301.005.

25. UN Commission on Human Rights, 'Summary Record of the Fifteenth Meeting [of the Drafting Committee of the Commission on Human Rights], 23 June 1947, E/CN.4/AC.1/SR.15,' 1947.

26. UN Commission on Human Rights, 'Summary Record of the Ninth Meeting [of the Working Group on the Declaration of Human Rights], 10 December 1947, E/CN.4/AC.2/SR.92,' 1947; UN Commission on Human Rights, 'Summary Record of the Seventieth Meeting [of the Commission on Human Rights], 11 June 1948, E/CN.4/SR.70,' 1948.

27. UN Commission on Human Rights, 'Summary Record of the Hundred and Fiftieth Meeting [of the Third Committee], 20 November 1948, A/C.3/SR.150,' 1948.

28. UN Commission on Human Rights, 'Summary Record of the Seventieth Meeting [of the Commission on Human Rights], 11 June 1948, E/CN.4/SR.70'; UN Commission on Human Rights, 'Summary Record of the Hundred and Fiftieth Meeting [of the Third Committee], 20 November 1948, A/C.3/SR.150.'

29. UN Commission on Human Rights, 'Summary Record of the Seventieth Meeting [of the Commission on Human Rights], 11 June 1948, E/CN.4/SR.70'; UN Commission on Human Rights, 'Summary Record of the Hundred and Fifty-Second Meeting [of the Third Committee], 22 November 1948, A/C.3/SR.152,' 1948.

30. UN Commission on Human Rights, 'Summary Record of the Hundred and Fiftieth Meeting [of the Third Committee], 20 November 1948, A/C.3/SR.150'; UN

Commission on Human Rights, 'Summary Record of the Hundred and Fifty-First Meeting [of the Third Committee], 22 November 1948, A/C.3/SR.151,' 1948; UN Commission on Human Rights, 'Summary Record of the Hundred and Fifty-Second Meeting [of the Third Committee], 22 November 1948, A/C.3/SR.152.'

31. UN Commission on Human Rights, 'Summary Record of the Two Hundred and Twenty-Eigth Meeting [of the Third Committee], 5 May 1951, A/C.3/SR.152,' 1951 Havet of UNESCO 'said that the principles applicable in that field were enunciated in Article 27 of the Universal Declaration of Human Rights.'

32. UN Commission on Human Rights, 'Summary Record of the Two Hundred and Ninety-Second Meeting [of the Third Committee], 13 May 1955, A/C.3/SR.292,' 1952; UN Commission on Human Rights, 'Summary Record of the Two Hundred and Ninety-Third Meeting [of the Third Committee], 14 May 1955, A/C.3/SR.293,' 1952; UN Commission on Human Rights, 'Summary Record of the Two Hundred and Ninety-Fourth Meeting [of the Third Committee], 14 May 1955, A/C.3/SR.294,' 1952; UN General Assembly, Third Committee, 'General Assembly Official Records, 12th Session: 3rd Committee, 796th Meeting, Thursday, 31 October 1957, New York, A/C.3/SR.796,' 1957; UN Commission on Human Rights, 'Summary Record of the Two Hundred and Twenty-Eigth Meeting [of the Third Committee], 5 May 1951, A/C.3/SR.152'; UN General Assembly, Third Committee, 'General Assembly Official Records, 12th Session: 3rd Committee, 799th Meeting, Thursday, 4 November 1957, New York, A/C.3/SR.799,' 1957.

33. Porsdam Mann, 'The Right to Science or to Wissenschaft?.'

34. The guarantee of scientific freedom in 15(3) is partially reflected in the word 'freely' used in Article 27 UDHR.

35. Porsdam Mann, 'The Right to Science or to Wissenschaft?.'

36. UN Commission on Human Rights, 'Summary Record of the Two Hundred and Twenty-Eighth Meeting [of the Third Committee], 5 May 1951, A/C.3/SR.152.'

37. UN Commission on Human Rights.

38. UNESCO, 'Constitution of the United Nations Educational, Scientific and Cultural Organization,' 1945 art 1(1).

39. UNESCO art 1(2).

40. Porsdam Mann, 'The Right to Science or to Wissenschaft?,' 21.

41. Porsdam Mann, 21.

42. 'Annotations on the Text of the Draft International Covenants on Human Rights (Prepared by the Secretary-General), A/2929,' 1955.

43. UN Commission on Human Rights, 'Summary Record of the Two Hundred and Ninety-Second Meeting [of the Third Committee], 13 May 1955, A/C.3/SR.292.'

44. UN Commission on Human Rights.

45. UN Commission on Human Rights.

46. UN Commission on Human Rights, 'Summary Record of the Two Hundred and Ninety-Third Meeting [of the Third Committee], 14 May 1955, A/C.3/SR.293.'

47. UN Commission on Human Rights, 'Summary Record of the Two Hundred and Ninety-Second Meeting [of the Third Committee], 13 May 1955, A/C.3/SR.292.'

48. UN Commission on Human Rights.

49. UN Commission on Human Rights.

50. UN General Assembly, Third Committee, 'General Assembly Official Records, 12th Session: 3rd Committee, 795th Meeting, Wednesday, 30 October 1957, New York, A/C.3/SR.795,' 1957 para. 14.

51. UN General Assembly, Third Committee, 'General Assembly Official Records, 12th Session: 3rd Committee, 796th Meeting, Thursday, 31 October 1957, New York, A/C.3/SR.796' para. 7.

52. UN General Assembly, Third Committee para. 22.

53. Memorandum by the Secretary-General, E/CN.4/673 (23 January 1953), 20. Paragraph 3 of Article 16 provides that 'The States Parties to the Covenant undertake to respect the freedom indispensable for scientific research and creative activity.' It is felt that the word 'indispensable' may be interpreted as limiting the freedom in question. The paragraph might be reworded as follows: 'The States Parties to the Covenant undertake to respect freedom of scientific research and creative activity.' Valuenzuela of Chile "urged the United States delegation to [. . .] delete the word 'necessary' in its own sub-paragraph (b), which in the present context might have a restrictive meaning.' E/CN.4/SR.292; Vela of Guatemala "said he approved of the original text of article 16 (E/2573, annex I A) but felt that in paragraph 3 the word 'indispensable' should be deleted because it could be interpreted in a restrictive sense; the same was true of the word 'necessary,'" which would not be any more satisfactory.' A/C.3/798, para. 28. Devasar of the Federation of Malaya's "delegation would support the Guatemalan representative's proposal that the word 'indispensable' in paragraph 3 of article 16 should be deleted. Subject to that, it would vote in favour of the article in its original form.' A/C.3/798, para. 49.

54. UN General Assembly, Third Committee, 'General Assembly Official Records, 12th Session: 3rd Committee, 796th Meeting, Thursday, 31 October 1957, New York, A/C.3/SR.796' para. 29.

55. UN General Assembly, Third Committee, 'General Assembly Official Records, 12th Session: 3rd Committee, 798th Meeting, Thursday, 1 November 1957, New York, A/C.3/SR.798,' 1957 para. 23.

56. The Philippines delegation twice requested a separate vote, at the 796th (para. 29) and 799th (para. 12) meetings. The Guatemalan delegation explicitly associated itself with this request at the 798th meeting (para. 28).

57. UN General Assembly, Third Committee, 'General Assembly Official Records, 12th Session: 3rd Committee, 799th Meeting, Thursday, 4 November 1957, New York, A/C.3/SR.799,' para. 18.

58. UN General Assembly, Third Committee.

59. UN Commission on Human Rights, 'Summary Record of the Two Hundred and Ninety-Second Meeting [of the Third Committee], 13 May 1955, A/C.3/SR.292.'

60. UN Commission on Human Rights.

61. UNESCO, 'Venice Statement on the Right to Enjoy the Benefits of Scientific Progress and Its Applications,' *SHS/RSP/HRS-GED/2009/PI/H/1*, 2009. Available at https://www.aaas.org/sites/default/files/VeniceStatement_July2009.pdf.

62. UNESCO, para 13.

63. UNESCO, 'Venice Statement on the Right to Enjoy the Benefits of Scientific Progress and Its Applications' para. 8.

64. UNESCO para. 14.

65. UNESCO para. 12(d).

66. UNESCO para. 14(c).

67. Farida Shaheed, 'The Right to Enjoy the Benefits of Scientific Progress and Its Applications, A/HRC/20/26,' 2012.

68. Shaheed, para. 41.

69. Shaheed para. 42.

70. Shaheed para. 47.

71. Shaheed para. 49.

72. UN Committee on Economic, Social and Cultural Rights, 'General Comment No. 25 (2020) on Science and Economic, Social and Cultural Rights, E/C.12/GC/25,' 2020.

73. UN Committee on Economic, Social and Cultural Rights para. 13.

74. UN Committee on Economic, Social and Cultural Rights para. 13.

75. UN Committee on Economic, Social and Cultural Rights para. 14.

76. See Chapter One of Part Three.

77. UN Committee on Economic, Social and Cultural Rights para. 15.

78. UN Committee on Economic, Social and Cultural Rights para. 43.

79. UN Committee on Economic, Social and Cultural Rights para. 46.

80. UN Committee on Economic, Social and Cultural Rights para. 22.

81. UN Committee on Economic, Social and Cultural Rights para. 52.

82. UN Committee on Economic, Social and Cultural Rights para. 57.

83. UN Committee on Economic, Social and Cultural Rights para. 55.

84. UN Committee on Economic, Social and Cultural Rights para. 20.

85. Christopher G. Weeramantry, *The Slumbering Sentinels: Law and Human Rights in the Wake of Technology* (London: Penguin Books, 1983).

86. Richard Pierre Claude, *Science in the Service of Human Rights* (Pennsylvania: University of Pennsylvania Press, 2002).

87. Audrey Chapman, 'A Human Rights Perspective on Intellectual Property'; Maria Green, 'Drafting History of the Article 15 (1) (c) of the International Covenant on Economic, Social and Cultural Rights,' *E/C12/2000/15*, 2000.

88. Lea Shaver, 'The Right to Science and Culture,' *Wisconsin Law Review* 2010, no. 1 (2010): 121–84; Audrey Chapman, 'Towards an Understanding of the Right to Enjoy the Benefits of Scientific Progress and Its Applications,' *Journal of Human Rights* 8, no. 1 (2009): 1–36.

89. Sebastian Porsdam Mann, Helle Porsdam, and Yvonne Donders, "Sleeping Beauty': The Right to Science as a Global Ethical Discourse,' *Human Rights Quarterly* 42, no. 2 (2020): 332–56.

90. Helle Porsdam and Sebastian Porsdam Mann, eds., *The Right to Science: Then and Now* (Cambridge: Cambridge University Press, 2022).

91. Porsdam Mann, Porsdam, and Donders, "Sleeping Beauty",

92. Rumiana Yotova and Bartha M. Knoppers, 'The Right to Benefit from Science and Its Implications for Genomic Data Sharing,' *European Journal of International Law* 31, no. 2 (1 September 2020): 665–91, https://doi.org/10.1093/ejil/chaa028.

93. United Nations, 'International Covenant on Economic, Social and Cultural Rights (Adopted 16 December 1966, Entered into Force 3 January 1976) UNGA Res 2200A (XXI) (ICESCR)' (n.d.) arts. 16–17.

94. Yotova and Knoppers, 'The Right to Benefit from Science,' 678.

95. Yotova and Knoppers, 685.

96. Cesare P. R. Romano and Andrea Boggio, 'Right to Science,' in *Max Planck Encyclopedia of Comparative Constitutional Law*, 2020.

97. Romano and Boggio.

98. Andrea Boggio, 'The Right to Participate In and Enjoy the Benefits of Scientific Progress and Its Applications: A Conceptual Map,' *New York International Law Review* 24, no. 2 (2021): 51.

99. Boggio, 51.

100. Klaus D. Beiter, 'Where Have All the Scientific and Academic Freedoms Gone? And What Is 'Adequate for Science'? The Right to Enjoy the Benefits of Scientific Progress and Its Applications,' *Israel Law Review* 52, no. 2 (July 2019): 233–91.

101. Johannes Morsink *The Universal Declaration of Human Rights: Origins, Drafting, and Intent* (Pennsylvania: University of Pennsylvania Press, 1999).

CHAPTER 4

1. 'UNESCO Recommendation on Science and Scientific Researchers,' especially point 5, in II, accessed march 16, 2023, https://unesdoc.unesco.org/ark:/48223/pf0000263618.locale=en.

2. 'UNESCO,' Preamble.

3. Mike Frick and Gisa Dang, 'The Right to Science: A Practical Tool for Advancing Global Health Equity and Promoting the Human Rights of People with Tuberculosis,' in *The Right to Science: Then and Now*, eds. Helle Porsdam and Sebastian Porsdam Mann (New York: Cambridge University Press, 2022), 246–267. This quotation may be found on p. 251. In their article, Frick and Dang link scientific development with equitable diffusion and argue that 'a state that satisfies its obligation to develop science without supporting diffusion and conservation has not created the conditions necessary for access' (Frick and Dang, 251).

4. See Helle Porsdam, *Science as a Cultural Human Right* (Cambridge University Press, 2022), chapter 6, for a discussion of this.

5. Gilbert Murray, 'International Organisation of Intellectual Work. Report Presented to the Assembly by the Fifth Committee' of the League of Nations, 1921, accessed on November 4, 2022, https://libraryresources.unog.ch/ld.php?content_id=31291797.

6. Preamble, 'The Covenant of the League of Nations,' League of Nations—Official Journal, February 1920, p. 3, accessed November 22, 2022, https://libraryresources.unog.ch/ld.php?content_id=32971179.

7. Albert Einstein, quoted in P. Speziali, 'Albert Einstein at the Commission on Intellectual Cooperation of the League of Nations,' *Europhys. News*, Volume 10, No.

4, 1979, 10–11, here at 10, accessed November 4, 2022, https://www.europhysicsnews
.org/articles/epn/pdf/1979/04/epn19791004p10.pdf.

8. Speziali, 10.

9. 'Centenary of the International Committee on Intellectual Cooperation of the League of Nations,' Conference in Geneva, May 12–13, 2022, accessed November 4, 2022, https://www.connections.clio-online.net/searching/id/event-116757?title =centenary-of-the-international-committee-on-intellectual-cooperation-of-the-league -of-nations&page=5&q=&sort=&fq=&total=7318&recno=92&subType=event &fbclid=IwAR3copWYhsKAgRVfEcsTx8TjWfNBAjCAAhduodVS1O7cTIzoOQ0 rCoTBVzw

10. 'Centenary of the International Committee.'

11. See e.g., Naomi Oreskes, *Why Trust Science?* (Princeton & Oxford: Princeton University Press, 2019).

12. Oreskes, pp. 49–51. This quotation can be found on p. 137.

13. Oreskes, pp. 68, 247.

14. CRISPR is an abbreviation of 'clustered regularly interspaced short palindromic repeats' and is a highly precise genetic engineering tool or method of editing and making changes to the base pairs of a gene.

15. Speaking about the American context, Naomi Oreskes offered these examples at the conference, 'Open World Conference 2022: Open Science and Global Dangers,' held at the University of Copenhagen on November 10 & 11, 2022 (www.openworld .ku.dk).

16. According to a Gallup's annual Governance survey, conducted in September 2022, 47% of U.S. adults say they have 'a great deal' or 'a fair amount' of trust in the judicial branch of the federal government. See Jeffrey M. Jones, 'Supreme Court Trust, Job Approval at Historical Lows,' *Politics*, September 29, 2022, accessed November 23, 2022, https://news.gallup.com/poll/402044/supreme-court-trust-job -approval-historical-lows.aspx. To this should be added that only 25% of Americans currently have confidence in the Court. This is down from 36% in 2021 and five percentage points lower than the previous low recorded in 2014. 'By all Gallup measures, then,' as Jeffrey M. Jones sums it up, 'Americans' opinions of the Supreme Court are the worst they have been in 50 years of polling.' See Jeffrey M. Jones, 'Confidence in U.S. Supreme Court Sinks to Historic Low,' *Politics*, June 23, 2022, accessed on November 23, 2022,

https://news.gallup.com/poll/394103/confidence-supreme-court-sinks-historic-low .aspx. This poll was conducted before the Court issued the *Dobbs v. Jackson Women's Health Organization* decision but after the leak of a draft opinion in that case signaled that the Court was about to overturn *Roe v. Wade.*

17. Thank you to Professor Peter Harder, University of Copenhagen, who shared his thoughts on the issue of scientific freedom with us at the launch of Helle Porsdam's book, *Science as a Cultural Human Right*, on Oct. 6, 2022.

18. Lincoln Caplan, 'Justice Elena Kagan, In Dissent: What to do about the loss of trust in the Supreme Court,' *Harvard Magazine*, November-December 2022, accessed November 4, 2022, https://www.harvardmagazine.com/2022/11/feature-justice-elena -kagan?utm_source=email&utm_medium=newsletter&utm_term=monthly&utm

_content=nd22&utm_campaign=101922. In the following few paragraphs, I rely on Caplan's article.

19. This ruling may well, Amy Howe argues, 'hamper President Joe Biden's plan to fight climate change and could limit the authority of federal agencies across the executive branch.' See Amy Howe, 'Supreme Court curtails EPA's authority to fight climate change,' Opinion Analysis, *SCOTUSblog*, June 30, 2022, accessed November 23, 2022 https://www.scotusblog.com/2022/06/supreme-court-curtails-epas-authority -to-fight-climate-change/.

20. Elena Kagan, quoted in Caplan, 'Justice Elena Kagan, In Dissent.'

21. Caplan.

22. Esther Marin-Gonzales et al., 'The Role of Dissemination as a Fundamental Part of a Research Project: Lessons Learned From SOPHIE,' *International Journal of Health Services* 47, no. 2 (2017): 258–276 https://doi.org/10.1177/002073141667622.

23. Gonzales et al.

24. Gonzales et al.

25. Susanne Beck, et al., 'The Value of Scientific Knowledge Dissemination for Scientists—A Value Capture Perspective,' *Publications* 7, no. 3(2019): 54; https://doi .org/10.3390/publications7030054

26. Beck, et al., 54.

27. Hans Peter Peters, 'Gap between science and media revisited: Scientists as public communicators,' *PNAS* 110, suppl. 3 (August 2013): 14102–14109, www.pnas .org/cgi/doi/10.1073/pnas.1212745110.

28. Beck et al., 'The Value of Scientific Knowledge Dissemination for Scientists.'

29. Beck et al.

30. Lucie White, 'A Neglected Ethical Issue in Citizen Science and DIY Biol-ogy,' *The American Journal of Bioethics* 19 no. 8, (2019): 46–48, https://doi.org/10 .1080/15265161.2019.1619876

31. David Sarpong et al., 'Do-it-yourself (DiY) science: The proliferation, rel-evance and concerns,' *Technological Forecasting and Social Change* 158 (2020) 120127, https://doi.org/10.1016/j.techfore.2020.120127.

32. Report of the Special Rapporteur in the field of cultural rights, A/HRC/20/26 (2012), para. 7.

33. Report of the Special Rapporteur, para. 43.

34. Report of the Special Rapporteur, para. 33.

35. Report of the Special Rapporteur, para. 52. I touch on these issues in both *The Transforming Power of Cultural Rights: A Promising Law and Humanities Approach* (Cambridge: Cambridge University Press, 2019) and *Science as a Cultural Human Right* (Pennsylvania: University of Pennsylvania Press, 2022).

36. Law concerning Danish universities, Articles 2, 3 – my translation from the Danish. The original says: 'Universitetet skal som central viden- og kulturbærende institution udveksle viden og kompetencer med det omgivende samfund og tilskynde medarbejderne til at deltage i den offentlige debat,' accessed December 11, 2022, https://www.retsinformation.dk/eli/lta/2019/778.

37. See *The Transforming Power of Cultural Rights*, pp. 151–53.

38. See e.g., U.S. National Science Foundation definitions of research categories, not dated, accessed December 11, 2022, https://www.radford.edu/content/dam/departments/administrative/sponsored-programs/PDFs/NSFdefinitions.pdf.

39. J. M. Santos, H. Horta & H. Luna, 'The relationship between academics' strategic research agendas and their preferences for basic research, applied research, or experimental development,' *Scientometrics* 127 (2022): 4191–4225, https://doi.org/10.1007/s11192-022-04431-5.

40. Kaare Aagaard, Maria Theresa Norn and Andreas Kjær Stage, 'How mission-driven policies challenge traditional research funding systems [version 1; peer review: awaiting peer review],' *F1000Research* 11, (2022): 949. https://doi.org/10.12688/f1000research.123367.1.

41. Aagaard, Norn and Stage.

42. Aagaard, Norn and Stage.

43. Aagaard, Norn and Stage.

44. Aagaard, Norn and Stage. For scholars motivated more by a wish to be engaged with applied research that addresses topical social issues and developments, or a wish to dedicate more time to knowledge transfer, this active public involvement will be perceived as less of a problem.

45. Hans Peter Peters, Abstract, 'Science Communication: Knowledge Dissemination or Public Engagement?' *Journal of Communication Research and Practice* 10 no. 1 (2020): 1–18 doi:10.6123/JCRP.202001_10(1).0001.

46. Hans Peter Peters.

47. Hans Peter Peters.

48. See Sebastian Porsdam Mann, 'The Right to Science or to Wissenschaft? A Chronology and Five Lessons from the Travaux Préparatoires,' *SSRN* (November 2022).

49. Incorporating 'inter-disciplinary and art and design elements in curricula and courses of all sciences as well as skills such as communication, leadership and management' as well as developing 'in each domain's curricula and courses the ethical dimensions of science and of research' is also what the 2017 UNESCO Recommendation on Science and Scientific Researchers recommends (Article 14, b+c).

50. UNESCO Chairs and UNITWIN Networks, https://www.unesco.org/en/education/unitwin.

51. UNESCO in brief, https://www.unesco.org/en/brief.

52. WIPO, 'Frequently Asked Questions: Copyright,' https://www.wipo.int/copyright/en/faq_copyright.html—accessed on 19 December 2022.

53. A/HRC/20/26 (2012), para. 57.

54. Please see Porsdam, *Science as a Cultural Human Right* for a more in-depth treatment of these issues.

55. United Nations 'Declaration of the Rights of Indigenous Peoples (UNDRIP),' https://www.un.org/development/desa/indigenouspeoples/wp-content/uploads/sites/19/2018/11/UNDRIP_E_web.pdf.

56. 'Charter of Fundamental Rights of the European Union, 2012/C 326/02,' Article 17,2: https://eur-lex.europa.eu/legal-content/EN/TXT/HTML/?uri=CELEX:12012P/TXT&from=EN.

57. AAAS, 'Defining the Right to Enjoy the Benefits of Scientific Progress and Its Applications: American Scientists' Perspectives' (Report prepared by Margaret Weigers Vitullo and Jessica Wyndham, Science and Human Rights Coalition), accessed December 20, 2022 https://www.aaas.org/sites/default/files/content_files/UNReportAAAS.pdf

58. AAAS, 6.

59. Christine Mitchel, 'Epilogue: Tensions in the Right to Science Then and Now,' in *The Right to Science: Then and Now* eds in Helle Porsdam and Sebastian Porsdam Mann (Cambridge: Cambridge University Press, 2019), 286–97. This quotation may be found at p. 294.

60. Tim Flink and Nicolas Rüffin, 'Chapter 6: The current state of the art of science diplomacy,' in *Handbook on Science and Public Policy* eds. Dagmar Simon, Stefan Kuhlmann, Julia Stamm and Weert Canzler (Cheltenham, UK: Edward Elgar Publishing, 2019)

61. Pierre-Bruno Ruffini, 'Conceptualizing science diplomacy in the practitioner-driven literature: a critical review,' *Humanities and Social Sciences Communications* 7, no. 124 (2020) https://doi.org/10.1057/s41599-020-00609-5. As promised in the title, Ruffino offers a critical assessment of the role of science diplomacy.

62. Finn Aaserud, 'Niels Bohr's Diplomatic Mission During and After World War Two,' *Berichte zur Wissenschaftsgeschichte* 43, no. 4 (2020): 493–520. https://doi.org/10.1002/bewi.202000026.

63. Niels Bohr, 'Open Letter to the United Nations,' 1950: http://www.atomicarchive.com/Docs/Deterrence/BohrUN.shtml.

64. Carlos Moedas, 'Science Diplomacy in the European Union,' *Science & Diplomacy*, Vol. 5, No. 1 (March 2016): http://www.sciencediplomacy.org/perspective/2016/science-diplomacy-in-european-union.

65. German Federal Foreign Office, 'Science Diplomacy: A new strategy in research and academic relations policy,' December 2020 – retrieved from https://www.auswaertiges-amt.de/blob/2436494/2b868e9f63a4f5ffe703faba680a61c0/201203-science-diplomacy-strategiepapier-data.pdf.

66. United Nations, Transforming our world: The 2030 Agenda for Sustainable Development, A/RES/70/1 (2015), https://sdgs.un.org/2030agenda; 2017 UNESCO 'Recommendation on Science and Scientific Researchers,' https://unesdoc.unesco.org/ark:/48223/pf0000260889.page=116; Committee on Economic, Social and Cultural Rights, 'General Comment No. 25 (7 April 2020) on Science and economic, social and cultural rights Art. 15.1.b, 15.2, 15.3 and 15.4,' E/C.12/GC/25.

67. AAAS, 'Defining the Right to Enjoy the Benefits of Scientific Progress and Its Applications,' 8.

68. AAAS, 8.

69. Marc Busch and David Logsdon, quoted in Patty Nieberg, 'The True Source of US-China Trade Tension is Over Technology,' *Medill News Service*, 4 February 2019, accessed January 1, 2023, https://dc.medill.northwestern.edu/blog/2019/02/04/the-true-source-of-u-s-china-trade-tension-is-over-technology/#sthash.4iZqhzbU.dpbs.

70. Andrew Silver, 'US-China trade war puts scientists in the cross hairs,' *Nature*, News, 22 June 2018, accessed December 31, 2022, https://www.nature.com/articles/d41586-018-05521-2.

71. CESCR, 'General Comment No. 25,' para. 54

72. Clare Wells, *The UN, UNESCO and the Politics of Knowledge* (New York: Macmillan Press, 1987), p. 170.

CHAPTER 5

1. United Nations, 'Vienna Convention on the Law of Treaties (Adopted 23 May 1969, Entry into Force 27 January 1980) 1155 UNTS 331 (VCLT)' (1969).

2. Sebastian Porsdam Mann, Yvonne Donders, and Helle Porsdam, 'The Right to Science in Practice: A Proposed Test in Four Stages,' in *The Right to Science: Then and Now*, ed. Helle Porsdam and Sebastian Porsdam Mann (Cambridge: Cambridge University Press, 2021), 231–45, https://doi.org/10.1017/9781108776301.005.

3. Philipp Alston, 'The Committee on Economic, Social and Cultural Rights,' in *The United Nations and Human Rights: A Critical Appraisal*, ed. Frédéric Mégret, 2nd ed. (Oxford University Press, 2020), 439–74.

4. Andrew Mazibrada, Monika Plozza, and Sebastian Porsdam Mann, 'Innovating in Uncharted Terrain: On Interpretation and Normative Legitimacy in the CESCR's General Comment No. 25 on the Right to Science,' *The International Journal of Human Rights*. In press (2023), https://papers.ssrn.com/abstract=4211453.

5. Mazibrada, Plozza and Porsdam Mann.

6. Alston, 'The Committee.'

7. UN Committee on Economic, Social and Cultural Rights (CESCR), "General Comment No. 25 (2020) on Science and Economic, Social and Cultural Rights, E/C.12/GC/25," 2020 para. 23; see also CESCR, "General Comment No. 3: The Nature of States Parties' Obligations (Art. 2, Para. 1, of the Covenant)," *E/1991/23*, 1991.

8. Amrei Müller, 'Limitations to and Derogations from Economic, Social and Cultural Rights,' *Human Rights Law Review* 9, no. 4 (1 January 2009): 557–601, https://doi.org/10.1093/hrlr/ngp027; Yvonne Donders, 'Balancing Interests: Limitations to the Right to Enjoy the Benefits of Scientific Progress and Its Applications,' *European Journal of Human Rights* October (1 October 2015): 486–503.

9. CESCR, 'General Comment No. 25 (2020) on Science and Economic, Social and Cultural Rights, E/C.12/GC/25,' para. 24.

10. CESCR, 'General Comment No. 25,' para. 25.

11. CESCR, 'General Comment No. 25,' para. 51.

12. CESCR, 'General Comment No. 25,' para. 52.

13. CESCR, 'General Comment No. 25,' para. 52.

14. CESCR, 'General Comment No. 3: The Nature of States Parties' Obligations (Art. 2, Para. 1, of the Covenant),' para. 13.

15. CESCR, 'General Comment No. 25 (2020) on Science and Economic, Social and Cultural Rights, E/C.12/GC/25,' para. 21.

16. UN Commission on Human Rights, "Note Verbale Dated 5 December 1986 from the Permanent Mission of the Netherlands to the United Nations Office at Geneva Addressed to the Centre for Human Rights ('Limburg Principles'), 8 January 1987, E/CN.4/1987/17,' 1987.

17. Müller, 'Limitations to and Derogations from Economic, Social and Cultural Rights.'

18. Müller.

19. Müller.

20. CESCR, 'General Comment No. 25 (2020) on Science and Economic, Social and Cultural Rights, E/C.12/GC/25,' para. 52.

21. Michael H. Andreae et al., 'An Ethical Exploration of Barriers to Research on Controlled Drugs,' *The American Journal of Bioethics* 16, no. 4 (2016): 36–47; CESCR, 'General Comment No. 25 (2020) on Science and Economic, Social and Cultural Rights, E/C.12/GC/25,' para. 68.

22. Andrew Mazibrada, Monika Plozza and Sebastion Porsdam Mann, 'Innovating in Uncharted Terrain: On Interpretation and Normative Legitimacy in the CESCR's General Comment No. 25 on the Right to Science,' *Organization Science* 31, no. 3 (2020): 535–57, https://doi.org/10.1287/orsc.2019.1328; Alston, 'The Committee,'; Gerd Oberleitner, "Understanding the Human Right to Science: CESCR General Comment No. 25 (2020)," in *The Human Rights-Based Approach to STEM Education*, ed. Tanja Tajmel, Klaus Starl, and Susanne Spintig (Münster: Waxmann, 2021).

23. Klaus D. Beiter, Terence Karran, and Kwadwo Appiagyei-Atua, 'Yearning to Belong: Finding a Home for the Right to Academic Freedom in the U.N. Human Rights Covenants,' *Intercultural Human Rights Law Review* 11 (2016): 107–90.

24. CESCR, 'General Comment No. 25,' para. 13.

25. CESCR, 'General Comment No. 25,' para. 46.

26. CESCR, 'General Comment No. 25,' para. 16.

27. CESCR, 'General Comment No. 25,' para. 26.

28. CESCR, 'General Comment No. 25,' para. 43.

29. CESCR, 'General Comment No. 25,' para. 45.

30. CESCR, 'General Comment No. 25,' para. 10.

31. UNESCO, 'Recommendation on Science and Scientific Researchers,' *SHS/BIO/PI/2017/3*, 2017.

32. James Mullin, 'Evaluation of Unesco's Regional Ministerial Conferences on the Application of Science and Technology to Development,' *126 EX/INF. 8*, 1987.

33. UNESCO, 'Desirability of Adopting an International Instrument on the Status of Scientific Research Workers,' *17 C/21*, 1971.

34. UNESCO, 'Recommendation on the Status of Scientific Researchers,' *18 C/Resolutions*, 1974.

35. UNESCO, 'Implementation of Standard-Setting Instruments,' *189 EX/13 Part III*, 2012.

36. UNESCO, 'Recommendation on Science and Scientific Researchers.'

37. UNESCO, 'Recommendation on Science and Scientific Researchers,' para. 19.

38. UNESCO, 'Recommendation on Science and Scientific Researchers,' 16(a)i.

39. UNESCO, 'Recommendation on Science and Scientific Researchers,' para. 10.

40. UNESCO, 'Recommendation on Science and Scientific Researchers,' para. 18b.

41. UNESCO, 'Recommendation on Science and Scientific Researchers,' para. 18(c).

42. UNESCO, 'Recommendation on Science and Scientific Researchers,' paras. 14, 27.

43. UNESCO, 'Recommendation on Science and Scientific Researchers,' para. 24.

44. UNESCO, 'Constitution of the United Nations Educational, Scientific and Cultural Organization,' 1945 preamble.

45. UNESCO/ICSU, 'Declaration on Science and the Use of Scientific Knowledge,' *30C/15/Annex I*, 1999.

46. UNESCO, 'Recommendation on Open Educational Resources (OER),' *CI-2022/WS/7 Rev.*, 2019.

47. UNESCO, 'Recommendation on Open Science,' *SC-PCB-SPP/2021/OS/UROS*, 2021 preamble.

48. UNESCO, 'Recommendation on Open Science,' para. 13(a).

49. UNESCO, 'Recommendation on Open Science,' para. 14(d).

50. UNESCO, 'Recommendation on Open Science,' para. 18.

51. UNESCO, 'Recommendation on Open Science,' para. 19.

52. UNESCO, 'Recommendation on Open Science,' para 17(f).

CHAPTER 6

1. UN Committee on Economic, Social and Cultural Rights (CESCR), 'General Comment No. 25 (2020) on Science and Economic, Social and Cultural Rights, E/C.12/GC/25,' 2020 para. 6. See generally on the this point Tara Smith, 'Scientific Purpose and Human Rights: Evaluating General Comment No 25 in Light of Major Discussions in the Travaux Préparatoires of the Universal Declaration of Human Rights and International Covenant on Economic, Social, and Cultural Rights,' *Nordic Journal of Human Rights* 38, no. 3 (2 July 2020): 221–36, https://doi.org/10.1080/18918131.2021.1882757.

2. CESCR, 'General Comment No. 25,' para. 37.

3. CESCR, 'General Comment No. 25,' para. 32.

4. UNESCO, 'Recommendation on Open Science,' *SC-PCB-SPP/2021/OS/UROS*, 2021 preamble.

5. UNESCO preamble.

6. Klaus D. Beiter, 'Where Have All the Scientific and Academic Freedoms Gone? And What Is 'Adequate for Science'? The Right to Enjoy the Benefits of Scientific Progress and Its Applications,' *Israel Law Review* 52, no. 2 (July 2019): 273.

7. Sebastian Porsdam Mann, 'The Right to Science or to Wissenschaft? A Chronology and Five Lessons from the Travaux Préparatoires,' *SSRN* (November 2022).

8. Stjepan Orešković and Sebastian Porsdam Mann, 'Science in the Times of SARS-CoV-2,' in *The Right to Science: Then and Now*, eds. Helle Porsdam and Sebastian Porsdam Mann (Cambridge: Cambridge University Press, 2021), 166–92, https://doi.org/10.1017/9781108776301.012; Sebastian Porsdam Mann et al.,

'UNESCO Brief on the Right to Science and COVID-19, SHS/IRD/2022/PI/1,' 2022, https://unesdoc.unesco.org/ark:/48223/pf0000381186.

9. UNESCO, 'Recommendation on Science and Scientific Researchers,' *SHS/BIO/PI/2017/3*, 2017 para. 6.

10. UNESCO, 'Recommendation on Science and Scientific Researchers,' para. 4.

11. UNESCO, 'Recommendation on Science and Scientific Researchers,' para. 16(1)(iv).

12. Sebastian Porsdam Mann et al., 'Is the Use of Modafinil, a Pharmacological Cognitive Enhancer, Cheating?,' *Ethics and Education* 13, no. 2 (4 May 2018): 251–67, https://doi.org/10.1080/17449642.2018.1443050.

13. Adam G. Dunn et al., 'Conflict of Interest Disclosure in Biomedical Research: A Review of Current Practices, Biases, and the Role of Public Registries in Improving Transparency' *Research Integrity and Peer Review* 1, no. 1 (3 May 2016): 1, https://doi.org/10.1186/s41073-016-0006-7.

14. Mariana Mazzucato, *Mission Economy* (London: Allen Lane, 2021).

15. UNESCO, 'Recommendation on Science and Scientific Researchers,' para. 10.

16. CESCR 'General Comment No. 25 (2020) on Science and Economic, Social and Cultural Rights, E/C.12/GC/25,' para. 55.

17. UNESCO, 'Recommendation on Open Science,' para. 14.

18. UNESCO, 'Recommendation on Science and Scientific Researchers,' para. 34(a).

19. UNESCO, 'Recommendation on Science and Scientific Researchers,' para. 34(c).

20. UNESCO, 'Recommendation on Open Science,' para. 20.

21. UNESCO, 'Recommendation on Open Science,' para. 20(c).

22. UNESCO, 'Recommendation on Open Science,' para. 20(d).

23. Margit Osterloh and Bruno S. Frey, 'How to Avoid Borrowed Plumes in Academia,' *Research Policy* 49, no. 1 (1 February 2020): 103831, https://doi.org/10.1016/j.rcspol.2019.103831.

24. Christopher R. Carpenter, David C. Cone, and Cathy C. Sarli, 'Using Publication Metrics to Highlight Academic Productivity and Research Impact,' *Academic Emergency Medicine* 21, no. 10 (2014): 1160–72, https://doi.org/10.1111/acem.12482.

25. Daniele Fanelli, 'Do Pressures to Publish Increase Scientists' Bias? An Empirical Support from US States Data,' *PLOS ONE* 5, no. 4 (21 April 2010): e10271, https://doi.org/10.1371/journal.pone.0010271.

26. Mark G. Siegel et al., "Publish or Perish' Promotes Medical Literature Quantity Over Quality,' *Arthroscopy: The Journal of Arthroscopic & Related Surgery* 34, no. 11 (1 November 2018): 2941–42, https://doi.org/10.1016/j.arthro.2018.08.029.

27. David Robert Grimes, Chris T. Bauch, and John P. A. Ioannidis, 'Modelling Science Trustworthiness under Publish or Perish Pressure,' *Royal Society Open Science* 5, no. 1 (10 January 2018): 171511, https://doi.org/10.1098/rsos.171511.

28. Mario Biagioli and Alexandra Lippman, eds., *Gaming the Metrics: Misconduct and Manipulation in Academic Research* (Massachusetts: MIT University Press, 2020).

29. Michael Park, Erin Leahey, and Russell J. Funk, 'Papers and Patents Are Becoming Less Disruptive over Time,' *Nature* 613, no. 7942 (January 2023): 138–44, https://doi.org/10.1038/s41586-022-05543-x.

30. CESCR, 'General Comment No. 25 (2020) on Science and Economic, Social and Cultural Rights, E/C.12/GC/25,' 25, para. 10.

31. CESCR, 'General Comment No. 25 (2020) on Science and Economic, Social and Cultural Rights,' para. 54.

32. UN Committee on Economic, Social and Cultural Rights paras. 55–56.

33. UNESCO, 'Recommendation on Open Science,' para. 10.

34. UNESCO, 'Recommendation on Open Science,' para. 21(e).

35. Sebastian Porsdam Mann, Helle Porsdam, and Yvonne Donders, "Sleeping Beauty': The Right to Science as a Global Ethical Discourse,' *Human Rights Quarterly* 42, no. 2 (2020): 332–56; see also Aurora Plomer, 'IP Rights and Human Rights: What History Tells Us and Why It Matters,' in *The Right to Science: Then and Now*, eds. Helle Porsdam and Sebastian Porsdam Mann (Cambridge: Cambridge University Press, 2021), 54–75, https://doi.org/10.1017/9781108776301.005.

36. UNESCO, 'Recommendation on Science and Scientific Researchers,' para. 16(b)iii.

37. UNESCO, 'Recommendation on Science and Scientific Researchers,' para. 37.

38. UNESCO, 'Recommendation on Science and Scientific Researchers,' para. 18(d).

39. CESCR, 'General Comment No. 25 (2020) on Science and Economic, Social and Cultural Rights,,' para. 60.

40. CESCR, 'General Comment No. 25 (2020) on Science and Economic, Social and Cultural Rights,,' para. 62.

41. UNESCO, 'Recommendation on Open Science' preamble.

42. UNESCO, 'Recommendation on Open Science,' para. 8.

43. Porsdam Mann, Porsdam, and Donders, "Sleeping Beauty.'

44. Beiter, 'Where Have All the Scientific and Academic Freedoms Gone?,' 277.

45. Ray Spier, 'The History of the Peer-Review Process,' *Trends in Biotechnology* 20, no. 8 (1 August 2002): 357–58, https://doi.org/10.1016/S0167-7799(02)01985-6.

46. Spier.

47. Samir Haffar, Fateh Bazerbachi, and M. Hassan Murad, 'Peer Review Bias: A Critical Review,' *Mayo Clinic Proceedings* 94, no. 4 (1 April 2019): 670–76, https://doi.org/10.1016/j.mayocp.2018.09.004.

48. Jonathan P Tennant, 'The State of the Art in Peer Review,' *FEMS Microbiology Letters* 365, no. 19 (1 October 2018): fny204, https://doi.org/10.1093/femsle/fny204.

49. Tennant.

50. Siladitya Jana, 'A History and Development of Peer-Review Process,' *Annals of Library and Information Studies (ALIS)* 66, no. 4 (7 December 2019), https://doi.org/10.56042/alis.v66i4.26964.

51. UNESCO, 'Recommendation on Science and Scientific Researchers,' para. 26.

52. UNESCO, 'Recommendation on Science and Scientific Researchers,' para. 1(a)i.

53. Tennant, 'The State of the Art in Peer Review.'

54. Tennant.

55. Haffar, Bazerbachi, and Murad, 'Peer Review Bias.'

56. UNESCO, 'Recommendation on Open Science,' para. 21(a).

57. UNESCO, 'Recommendation on Open Science,' para. 21(b).

58. CESCR, 'General Comment No. 25 (2020) on Science and Economic, Social and Cultural Rights,' para. 18.

59. Stephen Lock, 'Research Ethics—a Brief Historical Review to 1965,' *Journal of Internal Medicine* 238, no. 6 (December 1995): 513–20, https://doi.org/10.1111/j .1365-2796.1995.tb01234.x.

60. Lock.

61. Department of Health, Education, and Welfare and National Commission for the Protection of Human Subjects of Biomedical and Behavioral Research, 'The Belmont Report. Ethical Principles and Guidelines for the Protection of Human Subjects of Research,' *The Journal of the American College of Dentists* 81, no. 3 (2014): 4–13.

62. UNESCO, 'Recommendation on Science and Scientific Researchers,' para. 5(d).

63. CESCR, 'General Comment No. 25 (2020) on Science and Economic, Social and Cultural Rights,' para. 86.

64. CESCR, 'General Comment No. 25 (2020) on Science and Economic, Social and Cultural Rights,' para. 87.

65. CESCR, 'General Comment No. 25 (2020) on Science and Economic, Social and Cultural Rights,' paras. 19, 43.

66. Simon N. Whitney and Carl E. Schneider, 'Viewpoint: A Method to Estimate the Cost in Lives of Ethics Board Review of Biomedical Research,' *Journal of Internal Medicine* 269, no. 4 (April 2011): 396–402, https://doi.org/10.1111/j.1365-2796 .2011.02351_2.x.

67. Sebastian Porsdam Mann, Julian Savulescu, and Barbara J. Sahakian, 'Facilitating the Ethical Use of Health Data for the Benefit of Society: Electronic Health Records, Consent and the Duty of Easy Rescue,' *Philosophical Transactions. Series A, Mathematical, Physical, and Engineering Sciences* 374, no. 2083 (28 December 2016): 20160130, https://doi.org/10.1098/rsta.2016.0130; Sebastian Porsdam Mann et al., 'Blockchain, Consent and Prosent for Medical Research,' *Journal of Medical Ethics* 47, no. 4 (1 April 2021): 244–50, https://doi.org/10.1136/medethics-2019 -105963.

68. Robert Klitzman, *The Ethics Police? The Struggle to Make Human Research Safe* (Oxford: Oxford University Press, 2015).

69. Michael P. Diamond et al., 'The Efficiency of Single IRB Review in National Institute for Child Health and Human Development Cooperative Reproductive Medicine Network-Initiated Clinical Trials,' *Clinical Trials (London, England)* 16, no. 1 (February 2019): 3–10, https://doi.org/10.1177/1740774518807888.

70. Klitzman, *The Ethics Police?*

71. Nayha Sethi and Graeme T. Laurie, 'Delivering Proportionate Governance in the Era of EHealth: Making Linkage and Privacy Work Together,' *Medical Law International* 13, no. 2–3 (June 2013): 168–204, https://doi.org/10.1177/0968533213508974.

72. The 2017 Recommendation does not touch directly on the question of limitations.

73. UNESCO, 'Recommendation on Open Science,' para. 8. Emphasis added.

74. United Nations, 'International Covenant on Civil and Political Rights (Adopted 16 December 1966, Entered into Force 23 March 1976) UNGA Res 2200A (XXI) (ICESCR)' (n.d.), Art. 19(2–3).

75. Amrei Müller, 'Limitations to and Derogations from Economic, Social and Cultural Rights,' *Human Rights Law Review* 9, no. 4 (1 January 2009): 557–601, https://doi.org/10.1093/hrlr/ngp027.

76. Müller.

77. UNESCO, 'Recommendation on Science and Scientific Researchers,' para. 1(a)(i).

78. William Schabas, 'Study of the Right to Enjoy the Benefits of Scientific and Technological Progress and Its. Applications,' in *Human Rights in Education, Science and Culture: Legal Developments and Challenges*, ed. Yvonne Donders and Vladimir Volodin, n.d., 287; Amrei Müller, 'Remarks on the Venice Statement on the Right to Enjoy the Benefits of Scientific Progress and Its Applications (Article 15(1)(b) ICESCR),' *Human Rights Law Review* 10, no. 4 (1 December 2010): 765–84, https://doi.org/10.1093/hrlr/ngq033; Eide Riedel, 'Sleeping Beauty or Let Sleeping Dogs Lie? The Right of Everyone to Enjoy the Benefits of Scientific Progress and Its Applications (REBSPA),' in *Coexistence, Cooperation and Solidarity (2 Vols.)*, ed. Holger Hestermeyer et al. (Leiden: Brill/Nijhoff, 2012).

79. Marcos Orellana, 'Right to Science in the Context of Toxic Substances,' *A/HRC/48/61*, 2021.

80. See also Helle Porsdam, *Science as a Cultural Right* (Pennsylvania: University of Pennsylvania Press, 2022), Chapter 4 for a brief discussion of this issue.

81. The history of the emerging consensus definition of 'science' is discussed in much greater detail in ongoing thesis work by the first author. The thesis examines the scope and meaning of 'science' for the purposes of the right to science.

82. UNESCO, 'Science Policy in the European States,' *NS/SPS/25, SC.70/XIII.25/A*, 1971.

83. UNESCO, 'Science Policy in the European States,' para. 10.

84. UNESCO, 'International Instrument on the Status of Scientific Research Workers,' *SC/MD/35*, 1973, 5.

85. UNESCO, 'Recommendation on Open Science,' para. 4.

86. UNESCO, 'Recommendation on Open Science,' para. 6. Emphases added.

87. UNESCO, 'Recommendation on the Ethics of Artificial Intelligence,' *SHS/BIO/REC-AIETHICS/2021*, 2021, https://unesdoc.unesco.org/ark:/48223/pf0000380455. Emphases added.

CHAPTER 7

1. UN Committee on Economic, Social and Cultural Rights (CESCR), 'General Comment No. 25 (2020) on Science and Economic, Social and Cultural Rights, E/C.12/GC/25,' 2020.

2. See, among others, Helle Porsdam and Sebastian Porsdam Mann (eds.), *The Right to Science. Then and Now* (Cambridge: Cambridge University Press, 2022); Richard Pierre Claude, *Science in the Service of Human Rights* (Pennsylvania: University of Pennsylvania Press, 2011); William A. Schabas 'Study of the right to enjoy the benefits of scientific and technological progress and its applications,' in *Human rights in education, science and culture: legal developments and challenges*, eds. Donders, Yvonne and Volodin, Vladimir, (Oxfordshire: Routledge, 2007). Other essential texts about this right were written by Audry Chapman, Yvonne Donders, Lea Shaver, and Konstantinos Tararas. We must add Farida Shaheed's report as Special Rapporteur in the field of Cultural Rights (in 2012), and Margaret W. Vitullo and Jessica Wyndham's work for the American Association for the Advancement of Science (AAAS) among the pioneer milestones that paved the way.

3. At the 2009 Experts' Meeting in Venice, Yvonne Donders proposed four elements for the right to science: 1) scientific freedom; 2) the right to be protected from possible harmful effects of science; 3) access (including participation) in science; and 4) international cooperation. (The right to enjoy the benefits of scientific progress and its applications. Experts' Meeting on the Right to Enjoy the Benefits of Scientific Progress and its Applications, Venice. UNESCO, 2009).

4. Cesare P. R. Romano, 'The Origins of the Right to Science: The American Declaration on the Rights and Duties of Man,' in *The Right to Science: Then and Now*, eds. Helle Porsdam and Sebastian Porsdam Mann (Cambridge: Cambridge University Press, 2021), 33–53, https://doi.org/10.1017/9781108776301.004.

5. Mikel Mancisidor, 'The Dawning of a Right,' in *The Right to Science. Then and Now*, eds. Helle Porsdam and Sebastian Porsdam Mann (Cambridge: Cambridge University Press, 2022); Mikel Mancisidor, 'El derecho humano a la ciencia: Un viejo derecho con un gran futuro,' *Anuario de Derechos Humanos* 13 (2017): 211–221. For more general information see: Johannes Morsink, *The Universal Declaration of Human Rights: Origins, Drafting and Intent*, (Pennsylvania: University of Pennsylvania Press, 1999); William A. Schabas, *The Universal Declaration of Human Rights: The Travaux Préparatoires* (Cambridge: Cambridge University Press, 2017).

6. Richard Pierre Claude, *Science in the Service of Human Rights* (Pennsylvania: University of Pennsylvania Press, 2002); Sebastian Porsdam Mann, 'The Right to Science or to Wissenschaft? A Chronology and Five Lessons from the Travaux Préparatoires,' *SSRN* (November 2022).

7. The right to enjoy the benefits of scientific progress and its applications. Experts' Meeting on the Right to Enjoy the Benefits of Scientific Progress and its Applications, Venice, UNESCO, 2009.

8. Report of the Special Rapporteur in the Field of Cultural Rights, Farida Shaheed: 'The right to enjoy the benefits of scientific progress and its applications' (2012)

9. Shaheed, para. 25.

10. Shaheed, para. 75b.

11. UN Human Rights Council, Promotion of the enjoyment of the cultural rights of everyone and respect for cultural diversity, July 2012, A/HRC/20/L.18: '11. Recognises that further work and discussions on the issue are needed and, in that regard, requests the Office of the United Nations High Commissioner for Human Rights to

convene, in 2013, a seminar of two working days on the right to enjoy the benefits of scientific progress and its applications in order to further clarify the content and scope of this right and its relationship with other human rights and fundamental freedoms, including the right of everyone to the protection of the moral and material interests resulting from any scientific, literary or artistic production of which he or she is the author.'

12. United Nations High Commissioner for Human Rights, 'Report on the seminar on the right to enjoy the benefits of scientific progress and its application,' April 2014, A/HRC/26/19.

13. CESCR, 'Report on the fiftieth and fifty-first sessions (29 April–17 May 2013, 4–29 November 2013),E/2014/22 (E/C.12/2013/3)': '74. Regarding proposals for other general comments noted above, the Committee agreed to proposals from its members to carry out background research on article 15, paragraph 1 (b) on the right to enjoy the benefits of scientific progress and its applications, (rapporteurs Mr. Mancisidor and Mr. Marchán Romero); on State obligations in the context of corporate activities (rapporteur Mr. Kedzia) and on the pertinence of the Covenant rights to the environment and development (rapporteur Mr. Schrijver).'

14. CESCR, 'Report on the sixty-seventh and sixty-eighth sessions (17 February–6 March 2020, 28 September–16 October 2020), E/2021/22 E/C.12/2020/3': '68. At its sixty-seventh session, the Committee adopted general comment No. 25 (2020) on science and economic, social and cultural rights (article 15 (1) (b), (2), (3) and (4) of the Covenant). In the general comment, the Committee focuses on the Covenant right of everyone to enjoy the benefits of scientific progress and its applications (art. 15 (1) (b)) as a point of entry to analyse more broadly the relationship between science and all the Covenant rights.'

15. WHO Director-General's opening remarks at the media briefing on COVID-19, 11 March 2020, https://www.who.int/director-general/speeches/detail/who-director-general-s-opening-remarks-at-the-media-briefing-on-covid-19---11-march-2020.

16. United Nations High Commissioner for Human Rights, General discussion on a draft general comment on Article 15 of the ICESCR: on the right to enjoy the benefits of scientific progress, 9 October 2018, https://www.ohchr.org/en/events/days-general-discussion-dgd/2018/general-discussion-draft-general-comment-article-15-icescr.

17. In a very special way, I must thank the Diputación de Bizkaia (the government of the Territory of Bizkay) for its sustained support to the process for several years.

18. United Nations, 'What are General Comments of the Human Rights Treaty Bodies?,' 15 September 2021,https://ask.un.org/faq/135547#:~:text=A%20general%20comment%20is%20a,approaches%20to%20implementing%20treaty%20provisions.

19. As a practical example, in its Judgment (Sentencia) T-210 of 2018, the Colombian Constitutional Court ordered Colombian authorities to provide basic health services to Venezuelan migrants. One of the main arguments used by the Court was the General Comment No 14 on the Right to Health, https://www.corteconstitucional.gov.co/relatoria/2018/t-210-18.htm.

20. CESCR, 'General Comment No. 25 (2020) on Science and Economic, Social and Cultural Rights, E/C.12/GC/25,' para. 1.

21. Lea Shaver, 'The Right to Science and Culture,' *Wisconsin Law Review*, no. 1 (2010): 121–84.

22. Boutros Boutros-Ghali, 'The Right to Culture and the Universal Declaration of Human Rights' in *Cultural Rights as Human Rights*, ed. UNESCO (1970).

23. The Information Society Project's website can be accessed at: https://law.yale .edu/isp.

24. CESCR, 'General Comment No. 25 (2020) on Science and Economic, Social and Cultural Rights,' para. 11.

25. UNESCO, 'Recommendation on Science and Scientific Researchers,' *SHS/ BIO/PI/2017/3*, 2017.

26. CESCR, 'General Comment No. 25 (2020) on Science and Economic, Social and Cultural Rights,' para. 4.

27. CESCR, 'General Comment No. 25 (2020) on Science and Economic, Social and Cultural Rights,' para. 5.

28. CESCR, 'General Comment No. 25 (2020) on Science and Economic, Social and Cultural Rights,' para. 13.

29. CESCR, 'General Comment No. 25 (2020) on Science and Economic, Social and Cultural Rights,' para. 13.

30. CESCR, 'General Comment No. 25 (2020) on Science and Economic, Social and Cultural Rights,' para. 46.

31. CESCR, 'General Comment No. 25 (2020) on Science and Economic, Social and Cultural Rights,' para. 46.

32. Scholars at Risk, under the leadership of Robert Quinn, is developing a declaration of Principles Implementation of Academic Freedom that will be presented in the next months at the CESCR.

33. CESCR, 'General Comment No. 25 (2020) on Science and Economic, Social and Cultural Rights,' paras. 15–20.

34. CESCR, 'General Comment No. 25 (2020) on Science and Economic, Social and Cultural Rights,' para. 42.

35. CESCR, 'General Comment No. 25 (2020) on Science and Economic, Social and Cultural Rights,' paras. 43–44.

36. CESCR, 'General Comment No. 25 (2020) on Science and Economic, Social and Cultural Rights,' paras. 45–50.

37. CESCR, 'General Comment No. 25 (2020) on Science and Economic, Social and Cultural Rights,' para. 52.

38. Office of the High Commissioner for Human Rights, 'General Comment No. 15 (2002); The right to water (arts. 11 and 12 of the International Covenant on Economic, Social and Cultural Rights), E/C.12/2002/11.'

39. Outcome of the International Experts' Meeting on the Right to Water, Paris, 7 and 8 July 2009

40. See: Marisol Luna Leal and Nathaly Mendoza Zamudio, 'El derecho al agua en México. Algunas consideraciones,' *Lex Social* 5, no. 2 (2015).

41. Ministerio de Ciencia de España, https://www.ciencia.gob.es/Noticias /2021/Septiembre/La-ministra-Diana-Morant-reforzara-las-acciones-de-cultura -cientifica-con-el-mayor-presupuesto-en-15-anos.html; Ministerio de Educación y

Cultura de Uruguay, https://www.gub.uy/ministerio-educacion-cultura/comunicacion/convocatorias/disenar-derecho-ciencia.

42. In México, even though the proposal came before the adoption of the GC25 the wording of the proposal has links with the works of the CG ('consagrar a nivel constitucional el derecho a la ciencia, no es solamente el derecho a beneficiarse de los productos materiales de la ciencia y la tecnología; sino también, de sus aplicaciones; asimismo, es un derecho a beneficiarse del método científico y del conocimiento científico, sea para dar mayor capacidad a la toma de decisiones personales o para diseñar e implementar políticas públicas basadas en evidencia,' http://sil.gobernacion.gob.mx/Archivos/Documentos/2019/04/asun_3871352_20190429_1552421322.pdf); and in a much direct manner -and with the direct participation of a rapporteur of the GC25- in Chile, https://www.diarioconstitucional.cl/2022/06/08/unesco-coorganiza-coloquio-internacional-sobre-los-derechos-a-la-ciencia-y-sistemas-de-conocimiento-en-el-proceso-constituyente-chileno/; https://www.chileconvencion.cl/wp-content/uploads/2022/02/832-Iniciativa-Convencional-Constituyente-de-la-cc-Cristina-Dorador-sobre-Acceso-a-la-comunicacion-Cientifica.pdf; https://www.usach.cl/news/dr-cristian-parker-expuso-ante-comision-sistemas-conocimiento-la-convencion-constitucional.

43. UNESCO, 'Outcome of the International Experts' Meeting on the Right to Water, Paris, 7 and 8 July 2009,' https://unesdoc.unesco.org/ark:/48223/pf0000185432

44. UNESCO brief on the right to science and COVID-19 (SHS/IRD/2022/PI/1) 2022

45. 'This Brief draws attention to the human right to enjoy the benefits of scientific progress and its applications (referred to hereinafter as the 'right to science'), recognized in both Article 27(1) of the Universal Declaration of Human Rights (UDHR) and Article 15(1)b of the International Covenant on Economic, Social and Cultural Rights (ICESCR). This key human right, which underpins the vision of the UNESCO Recommendation on Science and Scientific Researchers (2017), is clarified by the CESCR General Comment No. 25,' UNESCO brief on the right to science and COVID-19, 3.

46. UNESCO, 'Join or MOOC on 'What does it mean to link science with human rights,' April 20, 2023, https://www.unesco.org/en/articles/join-our-mooc-what-does-it-mean-link-science-human-rights.

47. 'Diseñar el derecho a la ciencia,' November 11, 2021, https://es.unesco.org/sites/default/files/derciencia_basesconvocatoria.pdf

48. UNESCO, 'How to revive the right to science,' April 20, 2023, https://www.unesco.org/en/articles/how-revive-right-science

49. Secretariá General Iberoamericana, 'Declaración de Guatemala: COMPROMISO IBEROAMERICANO POR EL DESARROLLO SOSTENIBLE,' November 16, 2018, https://www.segib.org/wp-content/uploads/00.1.-DECLARACION-DE-LA-XXVI-CUMBRE-GUATEMALA_VF_E.pdf

50. OEI, 'Ciencia abierta y el derecho a la ciencia: antes, durante y post COVID-19,' October 5, 2020,
https://oei.int/pt/escritorios/uruguai/noticia/ciencia-abierta-y-el-derecho-a-la-ciencia-antes-durante-y-post-covid-19

CHAPTER 8

1. United Nations, 'Optional Protocol to the International Covenant on Economic, Social and Cultural Rights,' adopted 10 December 2008, entered into force 5 May 2013, 2922 UNTS 29.

2. United Nations, 'International Covenant on Economic, Social and Cultural Rights,' adopted 16 December 1966, entered into force 3 January 1976, 993 UNTS 3.

3. United Nations, 'International Covenant on Economic, Social and Cultural Rights,' accessed 30 August 2022, https://treaties.un.org/pages/ViewDetails.aspx?src =IND&mtdsg_no=IV-3&chapter=4&clang=_en.

4. United Nations, 'Optional Protocol to the International Covenant on Economic, Social and Cultural Rights,' accessed 30 August 2022, https://treaties.un.org/pages/ViewDetails.aspx?src=TREATY&mtdsg_no=IV-3-a&chapter=4.

5. https://juris.ohchr.org/en/search/results?Bodies=9&sortOrder=Date, last visited 30 August 2022.

6. *A.M.B. v. Ecuador*, CESCR 3/2014, 8 August 2016, E/C.12/58/D/3/2014, https://juris.ohchr.org/Search/Details/2137, last visited 30 August 2022.

7. *S.C and G.P. v. Italy*, CESCR 22/2017, 7 Mar 2019, E/C.12/65/D/22/2017, https://juris.ohchr.org/Search/Details/2522, last visited 30 August 2022.

8. Law No. 40 of 19 February 2004 'Rules on Medically Assisted Procreation' (in Italian, Norme in materia di procreazione medicalmente assistita), published in Official Gazette No 45, 24 February 2004, accessed 30 August 2022, https://www.gazzettaufficiale.it/eli/id/2004/02/24/004G0062/sg. A rough translation in English can be found in the website of the European Institute of Bioethics, accessed 30 August 2022, https://www.ieb-eib.org/docs/pdf/2010-12/doc-1554801408-32.pdf.

9. Italian Constitutional Court Judgment No. 151 of 8 May 2009, accessed 30 August 2022, http://www.cortecostituzionale.it/documenti/download/doc/recent_judgments/CC_SS_151_2009_EN.pdf,; Italian Constitutional Court Judgment No. 162 of 14 May 2014, accessed August 30, 2022, http://www.cortecostituzionale.it/documenti/download/doc/recent_judgments/162-2014_en.pdf; Italian Constitutional Court Judgment No. 96 of 14 May 2015, accessed August 30, 2022, http://www.cortecostituzionale.it/documenti/download/doc/recent_judgments/S96_2015_en.pdf; Italian Constitutional Court Judgment No. 229 of 21 October 2015, accessed August 30, 2022, https://www.cortecostituzionale.it/actionSchedaPronuncia.do?param_ecli=ECLI:IT:COST:2015:229.

10. *Costa and Pavan v Italy*, Application No. 54270/10, Eur. Ct. H.R., Judgment, 28 Aug. 2012, accessed August 30, 2022, http://hudoc.echr.coe.int/eng?i=001-112992; *Parrillo v. Italy*-46470/11 - Grand Chamber Judgment [2015] ECHR 755 (27 August 2015), accessed August 30, 2022, http://hudoc.echr.coe.int/eng?i=001-157263.

11. Andrea Boggio, 'Italy Enacts New Law on Medically Assisted Reproduction,' *Human Reproduction*, 20 (2005): 1153–57, https://academic.oup.com/humrep/article-lookup/doi/10.1093/humrep/deh871; John A. Robertson, 'Protecting Embryos and Burdening Women: Assisted Reproduction in Italy,' *Human Reproduction*, 19 (2004): 1693–96, https://academic.oup.com/humrep/article/19/8/1693/2356367

/Protecting-embryos-and-burdening-women-assisted; Vittorio Fineschi et al., 'The new Italian law on assisted reproduction technology (Law 40/2004),' *Journal of Medical Ethics,* 31, (2005): 536–39. For the story of the evolution, by litigation, of Law 40/2004, see Roberto Cippitani, 'The 'Curious Case' of Italian Law No. 40/2004: How the Dialogue between Judges is Modifying the Legislation on Medically-Assisted Reproduction,' *Rights and Science*, 0 (2017): 23–42; Rossana Cecchi et al., 'The law on artificial insemination: an Italian anomaly,' *Acta Biomed.*, 88 (2017): 403–8.

12. *S.C and G.P. v. Italy*, supra note 7, para. 2.1.
13. *S.C and G.P. v. Italy*, supra note 7, para. 2.1.
14. *S.C and G.P. v. Italy*, supra note 7, para. 2.3.
15. *S.C and G.P. v. Italy*, supra note 7, para. 2.3.
16. *S.C and G.P. v. Italy*, supra note 7, para. 2.3, note 2.
17. Italian Constitutional Court Judgment No. 151 of 8 May 2009, accessed 30 August 2022, http://www.cortecostituzionale.it/documenti/download/doc/recent_judgments/CC_SS_151_2009_EN.pdf.
18. *S.C and G.P. v. Italy*, supra note 7, para. 2.5.
19. *S.C and G.P. v. Italy*, supra note 7, para. 2.5.
20. *S.C and G.P. v. Italy*, supra note 7, para. 2.5.
21. *S.C and G.P. v. Italy*, supra note 7, para. 2.5.
22. *S.C and G.P. v. Italy*, supra note 7, para. 2.5.
23. *S.C and G.P. v. Italy*, supra note 7, para. 2.6.
24. *S.C and G.P. v. Italy*, supra note 7, para. 2.6.
25. *S.C and G.P. v. Italy*, supra note 7, para. 2.7.
26. *S.C and G.P. v. Italy*, supra note 7, para. 2.8.
27. Italian Constitutional Court Judgment N. 84/2016 of 22 March 2016, accessed August 30, 2022, https://www.cortecostituzionale.it/actionSchedaPronuncia.do?param_ecli=ECLI:IT:COST:2016:84.
28. Italian Constitutional Court Judgment N. 84/2016, para. 11.
29. *S.C and G.P. v. Italy*, supra note 7, para. 3.1–11.
30. *S.C and G.P. v. Italy*, supra note 7, para. 8.1–3.
31. *S.C and G.P. v. Italy*, supra note 7, para. 11.3.
32. *S.C and G.P. v. Italy*, supra note 7, para. 6.11–19.
33. *S.C and G.P. v. Italy*, supra note 7, para. 3.1.
34. *S.C and G.P. v. Italy*, supra note 7, para. 3.2.
35. *S.C and G.P. v. Italy*, supra note 7, para. 3.2.
36. *S.C and G.P. v. Italy*, supra note 7, para. 3.2.
37. *S.C and G.P. v. Italy*, supra note 7, para. 3.2.
38. *S.C and G.P. v. Italy*, supra note 7, para. 3.2.
39. *S.C and G.P. v. Italy*, supra note 7, para. 3.4.
40. *S.C and G.P. v. Italy*, supra note 7, para. 3.5.
41. *S.C and G.P. v. Italy*, supra note 7, para. 6.14.
42. *S.C and G.P. v. Italy*, supra note 7, para. 6.15.
43. *S.C and G.P. v. Italy*, supra note 7, para. 6.16.
44. *S.C and G.P. v. Italy*, supra note 7, para. 6.16.

45. *S.C and G.P. v. Italy*, supra note 7, para. 6.17.

46. *S.C and G.P. v. Italy*, supra note 7, para. 6.18.

47. CESCR, 'General Comment 25, on science and economic, social and cultural rights (Art. 15 (1)(b), (2), (3) and (4),' E/C.12/GC/25 (7 April 2020).

48. *S.C and G.P. v. Italy*, *supra* note 7, para. 6.16.

49. Rebecca Skloot, *The Immortal Life of Henrietta Lacks* (New York: Crown Publishing Group, 2010).

CHAPTER 9

1. Permanent Delegation of Denmark to UNESCO, 'Critical Voices—UNESCO's Instruments in Defence of Freedom of Expression of Artists, Journalists and Scientific Researchers.'

2. Bianca Nogrady, 'Scientists under attack' *Nature,* 598 (14 October 2021); Cathleen O'Grady, 'In the line of fire,' *Science* 375 (25 March 2022).

3. United Nations, 'Our Common Agenda'; Antonio Guterres, 'The Highest Aspiration—A Call to Action for Human Rights' by, UN Secretary-General on the occasion of the 75th anniversary of the UN, launched before the Human Rights Council in February 2020; United Nations, 'United Nations Guidance Note—Protection and Promotion of Civic Space,' September 2020.

4. UNESCO 'Brief on the Right to Science and COVID-19,' coordinated by UNESCO's Social and Human Sciences Sector, 2022; UNESCO 'Recommendation on Science and Scientific Researchers,' adopted by the General Conference of UNESCO, 13 November 2017

5. Statement by Nada Al-Nashif, UN Deputy High Commissioner for Human Rights, 50th session of the Human Rights Council, 22 June 2022, Geneva, Panel on Good Governance in Protecting Human Rights during and after COVID-19 pandemic. 'Raising the Ambition—Increasing the Pace,' UNGPs 10+, 'A Roadmap for the next decade of business and human rights,' published by UN Working Group on Business and Human Rights, Special Procedures, OHCHR, Geneva, November 2021; 'Guiding Principles on Business And Human Rights At 10: Taking stock of the first decade,' Geneva, June 2021.

6. Statement by Al-Nashif, 'Raising the Ambition—Increasing the Pace,' UNGPs 10+, 'A Roadmap for the next decade of business and human rights,' published by UN Working Group on Business and Human Rights; 'Guiding Principles on Business And Human Rights At 10: Taking stock of the first decade,' Geneva, June 2021.

7. Irene Khan, 'Reinforcing media freedom and the safety of journalists in the digital age,' Report of the Special Rapporteur on the promotion and protection of the right to freedom of opinion and expression, A/HRC/50/29, 20 April 2022.

8. United Nations, 'General Comment No. 25 concerning article 15 of the International Covenant on Economic, Social and Cultural Rights,' paras 13 and 52.

9. Human Rights Committee, 'Statement on derogations from the Covenant in connection with the COVID-19 pandemic,' CCPR/C/128/2, April 2020.

10. UNESCO 2017 'Recommendation on Science and Scientific Researchers,' para 16, (a), (i), para 16, (iii), paras 35–38.

11. O'Grady, 'In the line of fire.'

12. O'Grady.

13. Irene Khan, 'Reinforcing Media Freedom and the Safety of Journalists in the Digital Age, A/HRC/50/29,' para. 24.

14. United Nations, 'Guidance Note on the protection and promotion of civic space,' 2–3 and 10.

15. UNESCO, 'The United Nations Action Plan of Action on the Safety of Journalists and the Issue of Impunity.'

16. Khan, 'Reinforcing Media Freedom and the Safety of Journalists in the Digital Age, A/HRC/50/29'; United Nations, 'Guiding Principles on Business and Human Rights—Implementing the UN 'Protect, Respect and Remedy' Framework,' endorsed by the Human Rights Council in its resolution 17/4 of 16 June 2011.

17. Posetti et al., 'The Chilling: Global trends in online violence against women journalists,' discussion research paper, 2021, output of a wider UNESCO-commissioned global study, 'The Chilling: Assessing Big Tech's Response to Online Violence Against Women Journalists,' 2022, extracted chapter.

18. O'Grady, 'In the line of fire.'

19. UNESCO 'Recommendation,' paras. 16 (b)and 32; Posetti et al., 'The Chilling: What more can News Organisations do to Combat Gendered Online Violence?,' 2022, extracted chapter of the wider UNESCO-commissioned global study.

20. 177 ExB/Decision 35.I, and 196 ExB/Decision 20

21. 212 EX/23.III, paras. 6, 8, 10, 12, 20–22.

22. UNESCO, 'Preliminary Study on the technical and legal aspects relating to the desirability of revising the 1974 Recommendation on the Status of Scientific Researchers,' Revised First Draft, May 2013 (SHS/2013/PI/H/2 REV.2).

23. Valentina Carraro, 'Promoting Compliance with Human Rights: The Performance of the United Nations' Universal Periodic Review and Treaty Bodies' *International Studies Quarterly* 63, no. 4 (1 December 2019): 1087, https://doi.org/10.1093/isq/sqz078.

24. Permanent Delegation of Denmark to UNESCO, 'Critical Voices.'

25. Carraro, 'Promoting Compliance with Human Rights,' 1087.

26. Jane K. Cowan and Julie Billaud, 'Between learning and schooling: the politics of human rights monitoring at the Universal Periodic Review' *Third World Quarterly,* 36, no. 6 (June 2015): 1180–81, 1186–87.

27. Miloon Kothari, 'Study on emerging Good Practices from the Universal Periodic Review (UPR),' 12, https://www.ohchr.org/sites/default/files/Documents/HRBodies/UPR/Emerging_UPR_GoodPractices.pdf.

28. See database of accepted UPR recommendations related to scientific researchers, journalists, artists at unesco.um.dk.

29. Data, provided by the Danish Institute for Human Rights, is analysed in the report 'Critical Voices—UNESCO's Instruments in Defence of Freedom of Expression of Artists, Journalists and Scientific Researchers' produced by the Permanent Delegation of Denmark to UNESCO. The report is zooming in on those 2 per cent

illustrating key issues in relation to Member States' commitments to the freedom of expression for professional groups covered by UNESCO's mandate.

30. See 'Theme: D43 Freedom of opinion and expression' in the recommendations to Poland, available through the database of accepted UPR recommendations related to scientific researchers, journalists, artists at unesco.um.dk.

31. United Nations, 'General Comment No. 25 (2020) on science and economic, social and cultural rights (article (1) (b), (2), (3) and (4) of the International Covenant on Economic, Social and Cultural Rights),' para. 2.

32. Nordisk Kulturfond, 'The report 'Critical Voices' presented on the UN World Press Freedom Day,' May 19, 2022, https://www.nordiskkulturfond.org/en/news/the-report-critical-voices-presented-on-the-un-world-press-freedom-day.

33. Michael Randall, Learning Brief, 'Coalitions for Change—Collective action, better media ecosystems,' January 2022, https://www.mediasupport.org/publication/coalitions-for-change-collective-action-better-media-ecosystems/.

Index

About the Authors

Sebastian Porsdam Mann is a bioethicist and legal scholar. Trained in philosophy, biological and biomedical sciences and neuroethics at the University of Cambridge, he is currently pursuing a second doctoral degree in law at the University of Oxford. Sebastian has held postdoctoral positions at Harvard Medical School's Center for Bioethics and at the Universities of Copenhagen and Oxford, the latter funded by a Carlsberg Foundation grant. He has published widely on the human right to science and the ethical and human rights implications of emerging technologies like clinical proteomics, blockchain and generative artificial intelligence. Along with Helle Porsdam, he co-edited *The Right to Science: Then and Now*, published by Cambridge University Press in 2021.

Helle Porsdam is a professor of history and cultural rights and holds a UNESCO Chair at the University of Copenhagen. She earned her PhD at Yale University. Her research interests are American history and human rights, especially the right to science, and the role of science in society. In the fall of 2021, she was Leverhulme Visiting Professor at CRASSH, University of Cambridge. Her latest monograph, *Science as a Cultural Human Right*, was published in 2022 by the University of Pennsylvania Press (Studies in Human Rights).

Maximilian M. Schmid is the managing director of Cosana Europe and the Chief Executive Officer of A&BC Consulting. He holds degrees from the Munich and IE Business Schools. As an independent researcher, Maximilian has published on the ethical aspects of the human right to science and is currently pursuing work on the practical and ethical issues surrounding large-scale language models.

Péter V. Treit is a biochemical scientist specialising in metaproteomics of the human microbiome. He has published widely on proteomics, including on the ethical and societal aspects of clinical proteomics. Péter is currently pursuing a PhD at the Max-Planck Institute for Biochemistry in Munich, Germany.